[英国] 彼得·霍兰 著　王秀莉 译

# 动物

牛津通识读本·

# The Animal Kingdom

A Very Short Introduction

译林出版社

图书在版编目（CIP）数据

动物 /（英）彼得·霍兰（Peter Holland）著；王秀莉译.
—南京：译林出版社，2023.5
（牛津通识读本）
书名原文：The Animal Kingdom: A Very Short Introduction
ISBN 978-7-5447-9592-0

Ⅰ.①动… Ⅱ.①彼… ②王… Ⅲ.①动物－普及读物
Ⅳ.①Q95-49

中国国家版本馆 CIP 数据核字（2023）第 032965 号

*The Animal Kingdom: A Very Short Introduction* by Peter Holland

著作权合同登记号 图字：10-2017-080 号

**动物 ［英国］彼得·霍兰 / 著 王秀莉 / 译**

责任编辑 於 梅
装帧设计 韦 枫
校 对 戴小娥
责任印制 董 虎

原文出版 Oxford University Press, 2011
出版发行 译林出版社
地 址 南京市湖南路 1 号 A 楼
邮 箱 yilin@yilin.com
网 址 www.yilin.com
市场热线 025-86633278
排 版 南京展望文化发展有限公司
印 刷 江苏扬中印刷有限公司
开 本 890 毫米 × 1260 毫米 1/32
印 张 8.5
插 页 4
版 次 2023 年 5 月第 1 版
印 次 2023 年 5 月第 1 次印刷
书 号 ISBN 978-7-5447-9592-0
定 价 39.00 元

# 序 言

苗德岁

我们人类不仅是动物界的一员（即一个叫作“智人”的普通物种），而且与其他动物物种之间的关系至为密切，以至于说跟它们之间有着千丝万缕、唇齿相依和休戚与共的关系，一点儿也不为过。本书第五章的开篇引语“人类不过是蠕虫”，来自19世纪末英国幽默画刊《笨拙》上的一幅著名漫画的题图文字；这幅漫画是在达尔文生前最后一本书——《腐殖土的产生与蚯蚓的作用以及对其习性的观察》出版后，一位英国漫画家作来讽刺达尔文的生物演化论及人类起源于动物的“异端邪说”的：漫画中一条巨大的蚯蚓缠绕在达尔文的身上，让人看了忍俊不禁。

颇具讽刺意味的是，那幅画作发表140多年后，摆在我们面前的这本书，正是一本基于达尔文的生物演化论而撰写的极佳的动物学通识读物；读罢这本书，不得不想起已故著名遗传学家杜布赞斯基那句充满睿智的名言：“没有生物演化论，生物学里的一切都说不通。”

首先，什么是动物？如何给动物分类？世界上我们最熟视无睹的东西，往往最难定义，也最难分类；一如我曾经写过的："生命是什么？这个问题看似简单，其实至今都没有公认的答案，连生命科学家们对此也莫衷一是。我有时候想：大概生命就像爱情一样，似乎人人都知道它是什么，但是又很难给出一个严格的科学定义。"动物的定义也是如此，因而作者在本书第一章就开宗明义地专门讨论了"动物是什么"；其内容不仅新颖翔实、精彩丰富，而且堪称"惊艳"！如果我告诉你，变形虫（即阿米巴原虫）、领鞭虫以及其他许多"原生动物"不再算是动物，而你若是感到诧异的话，那么你便能在本书中找到"这是为什么"的答案！用多细胞以及上皮细胞层来定义动物，不仅现代、新颖、有趣，而且也反映了分子生物学与演化发育生物学的最新研究成果。这个定义使接下来有关动物起源的讨论，显得更加有的放矢、合情合理。可以毫不客气地说，单凭这些，过去的许多动物学教科书便可以送进废纸回收站了……

接下来的两章，作者深入讨论了动物的分类与演化。作者简要回顾了千百年来古代博物学家们与哲学家们探索动物分类与演化所走过的艰辛之路，从亚里士多德、邦纳、居维叶、拉马克、海克尔到达尔文和华莱士，人们眼见着"自然阶梯"的立起，又眼见着"自然阶梯"的倒塌……现代分子生物学、新达尔文主义和演化发育生物学相结合，才使我们对动物的分类与演化，有了全新和正确的认识；按照动物之间亲缘关系的相近程度对其进行分类，也才如实反映了动物的演化图景，因为亲缘关系相近表明它们之间有着血缘关系较近的共同祖先。本书作者把目前所知的动物按其相互的亲缘和演化关系分成了33个动物门，便于我们清晰地了解动物界各大类群的起源、机能、结构、演化和相互间的关系，而且他的论述极具趣味性。如果说160多年前达尔文在《物种起源》里提出的“生命之树”的概念是石破天惊的伟大发现的话（尽管还比较粗略），那么经过其后数代生命科学家们的不懈努力，本书作者已经成功地把这棵巨大的“生命之树”的枝叶和轮廓描绘得相当清晰、丰满和动人（见22页图2和119页图15）。

余下的第四至第十章，作者系统而简要地介绍了整个动物界从无脊椎动物到脊椎动物的各大类群的起源、演化与系统亲缘关系，在如此短小的篇幅里充满了如此丰富的硬核内容，这彰显了作者对基本材料和前沿研究的熟稔、内容剪裁的得体和叙述节奏的从容，整本书读来深入浅出、引人入胜。作者是知名分子遗传学家，在介绍和讨论“发育工具包”、古老基因组、Hox基因（同源异形基因）和演化发育系统生物学等方面的知识时，所用比喻形象、恰当、有趣，如描述细胞卵裂的排列时用堆橘子作比，将囊胚（肠道）的形成过程比喻成手指被推入一个冲了气的气球。这样的表述十分吸引人，给了我极大的阅读快感。显然，作者同时考虑到这是一本科普读物，为了引起一般读者的兴趣，还提到了许多与动物有关的趣闻逸事，比如有关七鳃鳗与欧洲王室的趣事（2006年我在《自然》上发表了七鳃鳗化石的论文后，曾应邀为《科学世界》写过一篇科普文章，其中有更详尽的相关叙述）。

最令我印象深刻的是，身为遗传学家的作者竟对古生物学研究成果了如指掌、应用自如。书中正确引述了许多古生物学的最新研究成果，并巧妙地结合了分子生物学领域的研究，对动物的起源与演化及系统关系，做了多学科交叉的精湛论述，实在令我惊叹不已！书中引用了罗默、古尔德、瓦伦丁、马古利斯、奥斯特姆、舒宾等我所熟知的众多古生物学界前辈或同行的论述，让我读来感到格外亲切。传统上，学术界也存在着“鄙视链”：伟大物理学家欧内斯特·卢瑟福的那句名言充满了偏见：“所有的科学，要么是物理学，要么是集邮活动。”他把物理学之外的所有学科（尤其是生命科学），视为集邮活动（不算是硬核科学）。在生命科学领域里，同样存在着鄙视链：遗传学家与古生物学家之间，向来也是相看两不“顺眼”的。我的学术前辈、著名古生物学家乔治·盖洛德·辛普森曾不无幽默地写道：“不久以前，古生物学家们感到遗传学家们只不过是把自己关在房间里，拉下窗帘，在牛奶瓶子里玩耍着小小的果蝇，却认为自己是在研究大自然！这种小把戏如此地脱离生命世界的现实，对真正的生物学家来说简直无足轻重。另一方面，遗传学家们则反唇相讥：古生物学家们除了证明演化真实发生之外，对生物学毫无建树、乏善可陈。古生物学不能算是真正的科学，古生物学家们像是一帮站在路旁看着汽车从身边飞驰而过，而试图去研究汽车发动机原理的人，可笑至极。”据说，他在哥伦比亚大学的同事、遗传学家杜布赞斯基在读到这段文字时，禁不住笑得前仰后合。说实话，即便是现在，我们与同系遗传学家同事之间的共同话题也不是很多。因此，我读这本书时，真是由衷地爱不释手。作者对古生物学的深度了解，令我肃然起敬！

最后一章既是对全书的精辟总结，也展望了未来动物学研究的愿景。本章的开篇引语是美国前国防部长拉姆斯菲尔德关于“已知与未知”的那段脍炙人口的名言，作者试图告诉读者，科学研究是无止境的，我们对“魅力无穷”的动物界虽然已知甚多，但未知更多。作者还列举了一些未来需要进一步探索的问题和方向。正如他在结尾所指出的："我相信，以动物学史的眼光来看，我们此时恰好第一次拥有了一棵可靠的动物多样性演化树。然而我们必须记住，这棵系统发生树只是生物学研究的起点……只有有了系统发生树的可靠框架，我们才能以有意义的方式比较动物物种之间的解剖构造、生理、行为、生态和发育，而这正是洞察生物演化模式和过程的路径。"我忍不住狗尾续貂："这也正是遗传学家们与古生物学家们携手合作，最可能做出突破性工作成果的康庄大道。"2022年诺贝尔生理学或医学奖颁发给了利用化石上残留的古DNA来研究人类演化历史的瑞典遗传学家斯万特·佩博，此举便充分说明了这一点。

总之，这是一本近年来我所读过的极为罕见的优秀动物学科普著作。尽管"牛津通识读本"系列中的众多书籍，从没有一本真正让我失望过，我不得不说，这是我所读过的最好的几本之一，并适合广泛的读者群——专家和外行读后，均会受益匪浅。而且，这是一本值得反复阅读的书，它将成为我的枕边书和案头书之一，甚至成为我的"沙漠或荒岛之书"的备选图书。

# 目 录

# 致　谢

本书的结构和内容，受益于牛津大学和雷丁大学过去及现在的学生。给苛刻而挑剔的学生讲授动物多样性课程迫使我就这门学科反复思考斟酌，学生的反馈也有助于突出关键问题。我要感谢牛津大学莫顿学院和牛津大学动物学系成员们的帮助，特别是西蒙·埃利斯和彭妮·申克的帮助。我还要感谢马克斯·特尔福德、克劳斯·尼尔森、比尔·麦金尼斯、斯图·韦斯特、特蕾莎·伯特·德佩雷拉、托比亚斯·乌勒、萨莉·利斯和佩尔·阿尔贝里等人对不同章节的指点，感谢塔季扬娜·索洛维约娃帮忙绘制了示意图。

i

第一章

# 动物是什么

我是现代将领的最佳楷模，

我熟知植物、动物和矿物。

——吉尔伯特和苏利文，

《潘赞斯的海盗》( 1879 )

## 动物的构成

根据我们的日常生活经验，判断哪些生物是动物，哪些不是，是轻而易举的。走在城镇里，我们会遇到猫、狗、鸟、蜗牛和蝴蝶，我们将所有这些都认定为动物。我们还应该把人类也加入动物名单中。同时，我们也会毫不迟疑地确定，我们看到的树木、花草和菌类不是动物，尽管它们也是生物有机体。而当我们开始考虑一些不那么常见的生物，其中很多是微生物的时候，定义或识别“动物”就变得困难了。因此，为回答“动物是什么？”这个问题而寻找精确标准是有帮助的。

所有动物共同拥有的一个特征是它们都是“多细胞的”。

也就是说，它们的身体由很多特化细胞构成。根据这一标准，单细胞生物，如为人熟知的阿米巴原虫，不被认为是动物，这与一个世纪前的观点相反。事实上，现在很多生物学家都谨慎地避免使用“原生动物”这个术语来指代阿米巴原虫这样的生物体，因为根据定义，一个生物体不可能既是“原生的”（意为“最初的”，即单细胞的），又是“动物”。

身体由多个细胞构成，这是一个必要条件，但仅此并不足够。同样的特性也存在于植物、真菌和黏菌等一些其他生物中，但它们都不是动物。动物的第二个重要特征是，它们获取生命所需能量的方式是吃掉其他的生物体，或生物体的一部分，这些生物体可能是死的，也可能是活的。这与绿色植物刚好形成对比，绿色植物利用太阳能在叶绿体内进行光合作用这一化学反应。有些植物将进食作为光合作用的补充（如捕蝇草），有些动物的体内有活的绿藻（如珊瑚和绿水螅），但这些基本不会模糊本质上的区别。



动物经常被提及的另一个特征是，动物具有移动和感知环境的能力。动物确实如此，但我们不能忘了，很多植物也有可以移动的部位，而细胞黏菌（并不是动物）可以形成一个缓缓移动的鼻涕虫形状的结构。

能产生精子和卵细胞是动物的另一个典型特征，虽然不同动物的精子和卵细胞大小相差很大。这一特征对动物行为的演化有着深远影响，但并不是一个能被轻易观察到的特征。也许只有仔细检查成年动物的细胞时才会发现高度一致的结构特征。尽管动物有很多不同类型的细胞，但有一类细胞影响了整个动物生物学和动物界的演化。这类细胞便是上皮细胞，它呈块状或长柱状，没有植物细胞那样的细胞壁。上皮细胞排列在一起形成柔韧的一片，相邻的细胞通过特殊的蛋白质连接在一起，还有其他的蛋白质封填细胞之间的缝隙，形成一个防水层。细胞层也存在于植物体内，但结构非常不同，而且韧性较小，更[2]容易透水。

无论是在功能上，还是在结构上，动物的上皮细胞都是极其重要的。上皮细胞可以控制上皮层两侧的液体的化学成分，容许动物创造出充满液体的空间，以支撑身体或汇集废物等。充满液体的空间是动物最早的骨架结构之一，促使了动物体格在演化过程中增大，同时也促进了动物的高效能运动。

此外，上皮细胞片由胶原蛋白等一层厚厚的蛋白质支撑，结实而柔韧，为精确的折叠运动提供了基础。这一特征在动物的胚胎发育过程中格外重要，折叠运动在发育过程中被用来生成动物身体的结构，就如同微型的折纸艺术一般。事实上，用纸来模仿动物发育的最初期阶段，是很简单的。尽管不同物种之间略有不同，但典型的动物发育都要经历一个阶段：上皮细胞形成小球（囊胚），这个小球是由一个细胞——受精卵——经历一系列的细胞分裂形成的。在大多数动物胚胎中，这个细胞球会在某个点或沿着一条凹槽向内折叠，将一些细胞移动到内部。之后形成一个管道，将来会发育为肠道，这是胚胎发育的关键步骤，名为原肠胚形成。而这个缩进去的球就叫作原肠胚。接下来还会发生进一步的折叠运动，形成充满液体的支撑结构、肌肉块，甚至脊髓和脑——就像在我们人类这样的脊椎动物体内一样。简而言之，细胞层构成了动物。

所有这些特征，都是我们用来识别动物的标准，我们也以此为基础，进一步了解动物的基本生活规律和现象。但这些并不能形成对动物最精准的定义。在生物分类学中，专家给演化树上或大或小的分支命名。一个基本规则是：真正的分组或“自然类群”所包含的生物群体，必须具有共同的演化祖先。这就意味着“动物”这个说法，必须指一组存在亲缘关系的物种。“动物”这个词，不能被用于演化树上其他位置的生物，即便它们具有一些类似动物的特征。同理，有一些物种，即便已经失去了它们的祖先所具有的正常的动物特征，我们依然会使用“动物”这个词来称呼它们。例如，一些动物在演化过程中失去了独特的精子和卵细胞，还有一些动物在其生命周期的每一个阶段都不是明显的多细胞生物，但是由于它们与其他动物有共同的祖先，所以被定义为动物。因此，动物是由同一个祖先演化而来的一个自然类群（或称分支）。这个分支被称为动物界，或称后生动物。



## 动物的起源

那个繁衍了所有动物的祖先在很久之前就已灭绝，但它是如何演化而来的呢？这似乎是一个很难解决的问题，因为我们要讨论的祖先已经灭绝了六亿年左右，它自然是极其微小的生物，而且没有留下化石记录。不过，令人惊讶的是，答案是相当确定的，且在140多年前就被首次提出。1866年，美国显微镜学家、哲学家、生物学家亨利·詹姆斯·克拉克指出，确定属于动物的海绵，其饲养细胞与一类鲜为人知的单细胞水生生物十分类似，这种生物当时被称作鞭毛纤毛虫。今天，我们称这类微生物为领鞭虫，DNA序列比较证明，它们确实是所有动物最近的

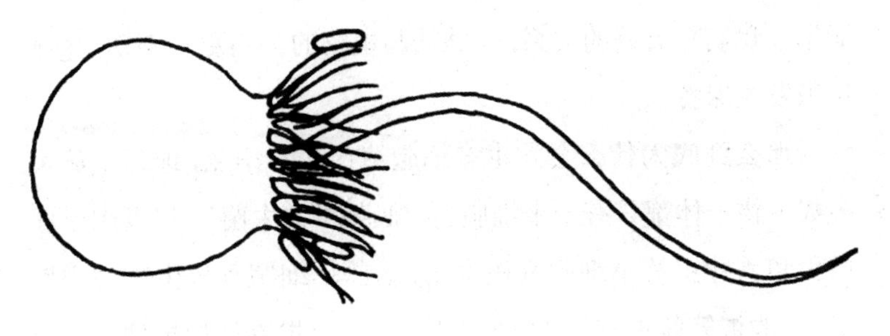

图1　领鞭虫，以细菌为食

近亲。领鞭虫和海绵的饲养细胞在一端都有一圈纤细的触须，这就是“领”，它们的形状就像是小小的羽毛球，不过在领圈的中央探出一根长长的鞭毛（一种可以移动的鞭状的结构）。领鞭虫的鞭毛在漂荡或击打时形成水流，食物微粒被带向细胞的方向，然后被领圈捕捉。海绵的饲养细胞运作方式与此不同，但同样使用鞭毛来形成水流。因此，所有动物最近代的祖先可能是一个微细胞球，其中的每个细胞都有一根鞭毛。动物界的起源涉及一系列变化，这些变化促使单细胞生命向微型水生细胞球转变。

4



请记住，动物并非地球上唯一的多细胞生物：植物、真菌、黏菌都是多细胞生命形式。然而，这些群体不是来自同一个祖先。它们是从不同的单细胞生物各自演化而来的。多细胞植物与动物（或领鞭虫）没有密切的亲缘关系，它们的演化过程构成生命树上一个完全不同的分支。真菌类如蘑菇、啤酒酵母和足癣，也和植物相去甚远，它们也是从自己的单细胞祖先演化而来的。下面这一说法可能令人惊讶：真菌及其祖先与动物和领鞭虫属于生命树的同一分支，一个被称作后鞭毛生物的群体。后鞭毛生物的多细胞演化发生了两次，一次动物诞生，另一次真菌诞生。我们与蘑菇的关系，比蘑菇与植物的关系更加密切，这个认识发人深省。

那么到底为什么会发生多细胞演化呢？毕竟，地球上绝大多数生物个体都只有一个细胞，如细菌、“古生菌”，以及阿米巴原虫和领鞭虫等单细胞真核生物，这些单细胞真核生物种类繁多，一般被笼统地称为“原生生物”。一个生物体如果拥有多个细胞，就可以长得更大，这便令其有能力避开其他细胞的捕食，或在单细胞生物无法适应的环境中定居。这么说也许没错，但这不太可能是多细胞演化的根本原因。毕竟，最初的多细胞生物，比如我们刚刚讨论的动物的祖先，可能也不过比微细胞球大一丁点而已，因而其栖息地和生活方式也受到限制，与其单细胞的近亲领鞭虫几乎相同。到目前为止，动物起源的问题还没有答案，不过有些很吸引人的想法已经被提了出来。一个独具巧思的观点是由林恩·马古利斯提出的，这个观点认为，多细胞结构使基本的分工成为可能：一些细胞能够分裂并生长，而另一些细胞依然可以继续摄食。但为什么单细胞不能在分裂的同时摄食呢？这个观点认为，像领鞭虫这样有鞭毛的单细胞生物，细胞组织的一个关键部分（被称作“微管组织中心”）要么在细胞分裂时移动染色体，要么掌控波动的鞭毛来摄食，但不能同时作用于两者。

还有一个富有想象力的模式，说来有点恐怖。这个观点是由米歇尔·凯尔斯伯格和刘易斯·沃尔珀特提出的，其核心理念为，以自体为食是多细胞演化的最初驱动力。假设有一个单细胞生物族群（如领鞭虫或它们的近亲）和一些细胞分裂后子细胞没有完全分离的突变体，这些突变体会形成细胞团，也就是细胞群落。单细胞生物与群落形式的突变体都从周围的水中过滤出细菌为食。当食物充足时，它们都可以获得营养并成功繁殖。然而，有些时候，也许由于环境的变化，食物会变得稀缺。许多生物会因为无法获得足够的营养来维持基本的细胞过程而死亡。然而，群落形式的突变体在困难时期有一个临时保障机制：细胞可以吃掉它们的邻居。这可以通过营养共享实现，群落中只有一部分细胞能存活下来，或者通过更戏剧性的过程实现：一些细胞衰变为邻近细胞的食物。因此，在食物短缺的时候，群落形式的突变体会有选择优势，更多的突变体能存活下来，并繁殖后代。自体消化听起来可能很可怕，但实际上这是很多动物在饥饿时使用的一种策略，从扁虫到人类皆是如此。

整个动物界都起源于这些古老的细胞群落。在过去的六亿年，甚至更长的时间里，这些细胞群落的后代通过演化而变得多样化并不断扩散，今天地球上千百万个不同的动物物种由此产生。动物起源于海洋，但后来，它们能占领淡水、陆地和天空，比如溪涧和河流中的扁虫和鱼，陆地上的蜗牛和蛇，以及空中的蝴蝶和鸟。一些动物如吸虫和绦虫，侵入其他动物的身体，而一些动物如海豚，又回到了海洋里。由于这种巨大的差异，动物的体形天差地别，尺度跨越极大。寄生的粘原虫和二胚虫变小并简化，体形和微细胞群落差不多大，而巨大的鲸则优雅地驾驭着它们100吨重的身体在海洋中纵横。为了理解这种巨大的多样性，我们需要重点讨论一下动物界最基本的分类单位：门。

第二章

# 动物门

> 分类是关于自然秩序的基础的理论，并不是仅仅为了避免混乱而编制的枯燥目录。
>
> ——斯蒂芬·杰伊·古尔德，
>
> 《奇妙生命》(1989)

## 模式与分支

几个世纪以来，博物学家和哲学家一直都在为了理解地球上生命的种类而努力。其中最早也最普遍的一个观点是“自然阶梯”，这个观点认为，有生命的东西——有时也包含无生命的东西——可以按照一个线性的等级排列。基于解剖学上的复杂性、宗教意义和实际用途等混杂形成的理念，梯子上的每一个上升的梯级都代表着逐级的“进步”。这个观点可以追溯到柏拉图和亚里士多德的思想，不过最早明确提出这一观点的，是18世纪瑞士博物学家查尔斯·邦纳的作品。在邦纳的体系中，“自然阶梯”以土和金属为起点，然后是石头和盐，再逐步发展，经过菌类、植物、海葵、蠕虫、昆虫、蜗牛、爬行动物、水蛇、鱼、鸟，最后至哺乳动物，人类在最顶端。或者说几乎是最顶端，还有天使和大天使略胜一筹。今天的我们要嘲笑这样的想法的确很容易，但邦纳对自然世界的认知是非常充分的，比如正是邦纳发现了蚜虫的无性繁殖，以及蝴蝶和它们的毛毛虫的呼吸方式。而且， 8
“自然阶梯”的概念在许多现代作品中仍然随处可见，许多科学家会谈论“高等”的或“低等”的动物：他们所使用的语言与这个古老而不可信的观点有着惊人的相似之处。

“自然阶梯”的倒塌，是逐步发生的。重大的一击来自备受尊敬的法国解剖学家、古生物学家、拿破仑的顾问居维叶男爵。根据对动物内部解剖的详细研究，居维叶得出结论：有四种本质上不同的构造身体的方式。这些不同不是表面上的差别，而是深深植根于神经系统、脑和血管的组织及功能的。1812年，居维叶将动物界分为四个分支：辐射动物（圆形的动物，如水母，令现代生物学家吃惊的是，其中还包括海星）、关节动物（身体可以分为一节一节的动物，如昆虫和蚯蚓）、软体动物（有壳和脑的动物）与脊椎动物（有骨骼、肌肉发达的心脏和红色血液的动物）。居维叶没有提出将这些分支联系起来的系统，因此，这些分支是平行发展的，地位平等，而非一个等级体系。

居维叶与同时代的拉马克不同，他不相信演化论。然而，矛盾的是，居维叶的分支能够以平等地位存在，背后的逻辑基础正是演化论提供的。正如后来查尔斯·达尔文和阿尔弗雷德·拉塞尔·华莱士都坚持的，演化论解释了为什么每个动物物种都与其他物种存在相似之处，以及为什么具有相同特征的物种群体可以被定义。将演化说成是一棵分支树，这是一个很常见的比喻，或者按照华莱士的更有诗意的说法，是一棵“多节的橡树”。利用这个比喻，我们可以把近亲物种的小“树枝”嵌入越来越大的分支，将相对较远的远亲囊括其中，所有这些物种都在演化中有着共同的祖先。然后，我们可以给这棵树上大大小小的分支起有意义的名字。动物界内大的分支被命名为“门”。

树的类比突出了动物分类系统的关键：命名必须反映演化产生的自然关系。给动物类群命名与给茶壶、邮票或啤酒杯垫这样的无生命的物体命名大不相同。无生命的物体可以根据不同的属性——如颜色、大小或原产国——分成多个不同的排列组合，所有这些属性都具有同等的相关性。但以这种方式对生物进行分类将会忽略一个基本要点：基于演化的分类系统反映了自然的秩序。这样的分类系统是一种关于亲缘关系的假说，是一种阐释独特的进化史的假说。



## 生命清单

有多少个动物门呢？换句话说，动物的演化树上有多少“大”的分支呢？这就引出了一个接下来的问题：一个分支必须有多大（或多小）才能被称为门。这是一个有争议的问题，但实际上，同一门的动物应该共同拥有不同于其他门动物的特殊解剖结构或特征。按照詹姆斯·瓦伦丁的说法：“门是生命树上基于形态学的分支。”一个门的名字永远不能用来给不同分支上的动物组合命名，一个门也不应该嵌套在另一个门里。这些规则能很好地适用于动物界的大部分动物，适用于所有非常为人熟悉的动物类型，但是对于需要多少门来对不太知名的物种进行分类，仍然存在争议和分歧。有一点毋庸置疑，居维叶的四个类别是过于粗略的简化归纳，今天我们经常提及的动物门的数量为30个至35个。

近年来，有几个“新”的门被提了出来。如果研究发现，一个门错误地包含了动物树上不同分支的动物，就会需要提出新的门，将这一个门分成两个门，这种情况偶有发生。一个例子是之前的中生动物门被分成了两个门：菱形动物门（其中包括一些小的蠕虫状寄生虫）和直泳动物门（其中包括一些更加微小的蠕虫状寄生虫，非常不可思议的是，这些动物生活在章鱼和鱿鱼的尿液中）。一个更有争议的例子是扁形动物门（如扁虫、绦虫和吸虫），近年来，其中的一些物种被移出，并被放入一个新的门——无腔动物门。如果发现了具有不寻常且明显独特的身体结构的全新物种，并且这一物种无法归入一个既有的门中，也会提出新的门。要确立一个新的门，两个标准都必须满足。自20世纪80年代以来，这种情况只发生过寥寥几次，比较知名的是环口动物门（附生在龙虾和海螯虾口器上的小动物）、铠甲动物门（附着在沙粒上的微型瓮状动物）和微颚动物门（在格陵兰的一处淡水泉中发现的更小的动物）的发现。

门也会消失。这并非由于物种灭绝，至少我们可以说，自从有人类记录以来，没有任何一个门灭绝。门被取消，是因为发现整个群体实际上属于另一个门。从逻辑上讲，这两个群体必须合二为一。出人意料的是，这种情况经常发生，最常见的情况是，一群解剖结构非常奇怪的动物最初被视为一个独立的门，但后来的研究发现，它们实际上是另一类动物的成员。最好的例子是巨型管虫，或称须腕动物，它们是生活在加拉帕戈斯群岛和大西洋中脊附近深海热液喷口处的著名居民。某些须腕动物可以生长到体长两米，但令人吃惊的是，它们的演化关系很难追踪。不过，DNA序列数据显示，须腕动物应该被认定为环节动物门的演化成员，环节动物门中包含蚯蚓和水蛭等众所周知的动物。另一个例子是曾经存在的五口动物门，又称舌形动物门，其中包括钩在鸟类和爬行动物鼻腔通道内的大型有鳞寄生虫（长达15厘米）。尽管外表恐怖，但DNA和细胞结构的分析表明，舌形虫实际上是高度变形的甲壳纲动物，完全可以放在节肢动物门下，和鱼虱是近亲。

我在本书中确认了33个动物门。其中的九个包含了几乎每个人都熟知的动物。还有四个，如果稍微花些力气，就能够在池塘、沟渠里或在海边漫步时看到。这九个广为人知的动物门为：多孔动物门（海绵）、刺胞动物门（水母、珊瑚、海葵）、节肢动物门（包括昆虫、蜘蛛、螃蟹、蜈蚣）、线虫动物门（导致河盲症的人体内的寄生虫蛔虫、园丁用来杀死蛞蝓的线虫）、环节动物门（蚯蚓、沙蚕、水蛭）、软体动物门（蜗牛、牡蛎、章鱼）、扁形动物门（扁虫、吸虫、绦虫）、棘皮动物门（海星、海胆）和脊索动物门（鱼、青蛙、蜥蜴、鸟以及包括人在内的哺乳动物）。另外四个比较容易发现的门是苔藓动物门（很容易被看到的是海藻叶上一排排微小的砖形苔藓动物）、纽形动物门（能在海滩的石头下找到软软的、缓缓移动的纽虫）、轮虫动物门（池塘中生活着许许多多的轮虫）和缓步动物门（能在苔藓中发现微型的“水熊”，位列大多数动物学家可爱动物名单的榜首）。

要理解动物如何通过演化而变得多样化，它们如何发挥机能，以及它们如何适应特定环境，最好从门这一层级开始。由于门是“生命树上基于形态的分支”，知道一个物种属于哪一个门，有助于我们将其与同一门内的其他物种或相近门内的物种进行比较，并思考该生物的解剖结构与身体机能之间的关系。比如知道一种动物属于线虫动物门，会因此立刻关注到这一门中发现的有弹性的厚表皮和咽泵，它们有助于理解该动物的生活方式和特性。相反，忽视分类会导致对远亲物种进行令人困惑的比较，这些物种可能拥有非常不同的身体构造，以及演化和生活方式方面的不同限制。但我们不能将动物门简单地理解为33个不同的分类。它们各自构成了演化树上的一个分支。当然，分支之间总是相互联系的。因此，一些门彼此之间的关系比它们与其他门之间的关系要密切。记住这一点至关重要，它有助于我们理解动物界的结构、机能和演化是如何相互联系的。

12



**表1**

| 门 | 演化树上的位置 | 代表物种 |
|---|---|---|
| 扁盘动物门 | 基础动物 | |
| 多孔动物门 | 基础动物 | 海绵 |
| 刺胞动物门 | 基础动物 | 水母、珊瑚、海葵 |
| 栉水母动物门 | 基础动物 | 栉水母 |
| 环节动物门 | 冠轮动物总门 | 蚯蚓、沙蚕、水蛭 |
| 软体动物门 | 冠轮动物总门 | 蜗牛、牡蛎、鱿鱼、章鱼 |
| 纽形动物门 | 冠轮动物总门 | 纽虫 |
| 腕足动物门 | 冠轮动物总门 | 腕足动物 |
| 帚虫动物门 | 冠轮动物总门 | 帚虫 |
| 苔藓动物门 | 冠轮动物总门 | 苔藓动物 |
| 内肛动物门 | 冠轮动物总门 | |

13

续 表

| 门 | 演化树上的位置 | 代表物种 |
| --- | --- | --- |
| 扁形动物门 | 冠轮动物总门 | 扁虫、吸虫、绦虫 |
| 二胚虫动物门 | 冠轮动物总门 | |
| 轮虫动物门 | 冠轮动物总门 | 轮虫 |
| 腹毛动物门 | 冠轮动物总门 | |
| 颚胃动物门 | 冠轮动物总门 | |
| 微颚动物门 | 冠轮动物总门 | |
| 环口动物门 | 冠轮动物总门 | |
| 节肢动物门 | 蜕皮动物总门 | 昆虫、蜘蛛、螃蟹、蜈蚣 |
| 有爪动物门 | 蜕皮动物总门 | 天鹅绒虫 |
| 缓步动物门 | 蜕皮动物总门 | 水熊 |
| 线虫动物门 | 蜕皮动物总门 | 蛔虫 |
| 线形动物门 | 蜕皮动物总门 | 马毛虫 |
| 动吻动物门 | 蜕皮动物总门 | 动吻虫 |
| 鳃曳动物门 | 蜕皮动物总门 | 鳃曳虫 |
| 铠甲动物门 | 蜕皮动物总门 | |
| 棘皮动物门 | 后口动物总门 | 海星、海胆、海参 |
| 半索动物门 | 后口动物总门 | 柱头虫 |
| 脊索动物门 | 后口动物总门 | 海鞘、文昌鱼、鱼、人 |
| 毛颚动物门 | 冠轮动物总门/<br>蜕皮动物总门 | 箭虫 |
| 无腔动物门 | 不确定 | |
| 异涡动物门 | 不确定 | |
| 直泳动物门 | 不确定 | |

第三章

# 动物演化树

> 尽管我不能活着看到，但我相信，总有一日，我们将拥有自然界每个伟大王国的非常真实的家谱树。
>
> ——查尔斯·达尔文，
>
> 《致T. H.赫胥黎的信》(1857)

## 构建一棵生命树

达尔文意识到，要描述演化过程，分支树是一个很好的比喻。1837年，他在一本私人笔记本上画了一幅演化树的小草图，并写下了诱人的字眼："我认为。"一个物种可以产生两个或更多的"子"物种——这个过程被称为物种形成，一旦达尔文意识到这个过程，生命树的概念很快就出现在他的脑海中。演化树，或称"系统发生树"，是描述这些物种形成事件的简单图表。系统发生树上的每一个分支点，每一条线变成两条线的地方，都形象地展示出一个物种变成了两个物种。

包含相似的动物物种时，系统发生树很容易理解。例如，如果树上的一条线通向欧洲粉蝶，另一条线通向菜粉蝶，这两条线的交点便标志着这两种非常相似的蝴蝶分化的物种形成事件。[15]这个历史上的点，便是两个有着“共同祖先”的种群从此分开，不能再异种交配的点。重要的是，在这个点上，这两个种群还没有获得两个物种的明显特征；实际上，它们看起来可能本质上是一样的。但通常来说，系统发生树并不只包含近亲物种，它们会描绘大的动物族群之间的演化关系，比如昆虫、蜘蛛、蜗牛、水母和人类之间的演化关系。但看待这些树的方式应该完全相同。如果树上的一条线通向昆虫，另一条线通向蜘蛛，那么这两条线相交的地方，标志着这两类动物早已灭绝的共同祖先。这个祖先既不是昆虫，也不是蜘蛛，在物种形成的过程中，这两类动物的祖先产生了。



尽管达尔文在他的笔记本上勾画了演化树的轮廓，并在《物种起源》唯一的一幅插图中将这棵树放大，但他并没有试图去解答谁和谁实际上是亲戚。对达尔文来说，演化树还仅仅是一个概念，一种对演化的思考方式。在他之后，很多演化生物学家都试图将物种的名字放在树的分支上。这是一个重要的问题，也是一个需要解决的问题。毕竟，应该有一棵单独的动物生命树，描绘动物演化的真实过程。因此，所有系统发生树的绘制，都是对演化路径清晰而明确的假设。德国动物学家恩斯特·海克尔于19世纪六七十年代绘制了最早的演化树。他笔下的许多树的艺术细节都引人注目，包括生着节疤的树皮和弯弯曲曲的树枝，每根小树枝和每片树叶的末端都标着特定动物类别的名字。海克尔的树和关于动物演化的假设，基于数条证据链，不过他特别喜欢将胚胎学上的特点作为证据，部分是因为他认为胚胎在演化过程中变化缓慢，即便长成后看起来大相径庭，但能在它们的发育过程中发现相似的特征。海克尔的一些结论和现代的一些观点也是一致的，比如他将水母和海葵放在了一个与其他动物生命很早就分开的分支上。他的一些其他想法在我们今天看来十分奇怪，而且当然是不正确的，比如将棘皮动物（海星和海胆）看作与昆虫、蜘蛛等节肢动物非常近的一个分支。

16

在接下来的80多年中，动物学家们对动物解剖学有了更好的阐释，更详细地研究了它们的胚胎发育，并格外关注无脊椎动物门的巨大多样性。但即便到了20世纪中叶，也还没有达成明确的共识。动物界中一直都没有一棵被普遍接受的系统发生树。每一位绘者都会给出一棵稍微有些不同的演化树，尽管某些特定的关系是一直存在的。我们下面要介绍的这种设想，在美国的教科书中十分流行，被称作“体腔演化假说”。

## 体腔演化假说

在这棵演化树中，用以决定哪些动物门关系最近的主要证据链是对称性、胚层、体腔、分节以及胚胎发育早期细胞分裂的模式。大多数熟悉的动物，包括蠕虫、蜗牛、昆虫和我们人类，都只有一个镜像平面，也就是说只有一条对称轴。这条对称轴从头向尾，将身体的左右两侧分开，而两侧呈镜像对称。有许多偏离精确对称的情况，比如蜗牛的螺旋壳、螃蟹不对称的钳子，或者人类在身体左侧的心脏，但这些都是微小的改变。从根本上说，大多数动物都有近乎镜像的左右两侧——一种被称为“两侧对称”的组织结构。有四个动物门与此不同，没有明显的头端和尾端，没有办法区分左边或右边。这几个“非两侧对称”的门，被称作基础动物门，门内动物要么没有对称性，要么呈辐射对称，包括刺胞动物门（水母、海葵、珊瑚）和多孔动物门（海绵），另外还有两个不太为人知的类别，名为栉水母动物门（栉水母）和扁盘动物门。17

第二条证据链是胚层的数目。胚层是胚胎早期出现的细胞层，会在发育过程中变得复杂。大多数动物有三层胚层，内层（内胚层）发育成肠壁，外层（外胚层）发育成皮肤和神经，中间层（中胚层）发育成肌肉、血液和其他组织。然而，非两侧对称的基础动物门只有两个胚层（外胚层和内胚层），至少根据最初的估计是如此。这些动物是否有中胚层的类似物是有争议的。由于对称和胚层这两条证据链，两侧对称的动物被放在一个大的类别中，被称为“两侧对称动物”（也因为拥有三个胚层而被称作“三胚层动物”），这一类别在动物演化早期就形成了不同分支。

接下来讨论两侧对称动物。在体腔系统发生理论中被特别关注的一个特征，是体内是否存在充满液体的空间。一些两侧对称动物，最有代表性的是环节动物（如蚯蚓）和软体动物（如蛞蝓和蜗牛）的胚胎，都有很大的充满液体的体腔，腔内有防水的上皮细胞层。脊索动物（如人类）和棘皮动物（如海星和海胆）的胚胎都有这些空腔。身体中的这种腔被称作体腔，这些动物因此被称作体腔动物，它们在演化树上是彼此靠近的群体。蚯蚓的体腔一直延续到成年，充当液体骨架。其他一些动物，包括节肢动物（如昆虫和蜘蛛）的体腔在发育后期可能会变得很小或消失，但是这些动物依然被放在演化树的体腔动物部分。（在其他一些系统发生树中，只有一部分体腔动物是聚集在一起的。）节肢动物被放在环节动物附近的另一个原因是，这两种动物的身体都被分成重复的关节单元，也就是“节”。在蜈蚣或蚯蚓的身体上，可以很清楚地看出这种节是身体周围（或内部）一系列的环。因此，在很多的演化树上，都专门定义了一个“分节的体腔动物”的超级分支，这个分支被称作关节动物。 18

与体腔动物形成对比的是，两侧对称动物中也有一些动物的中胚层保持固态，没有充满液体的体腔。这些动物被称作无体腔动物，其中包括扁形动物门（扁虫、吸虫、绦虫）和纽形动物门（纽虫）。介于这两个大分类之间的，是假体腔动物，如线虫动物门（蛔虫）。假体腔动物没有上皮细胞层，因而没有界限清晰的体腔。体腔发育假说认为所有的体腔动物是一个大类，它与无体腔动物在较早的阶段就分开演化了。无体腔动物被认为是体腔动物的祖先，因此是最“原始”的两侧对称动物。此外，这还意味着在两侧对称动物的演化过程中，从无体腔动物到体腔动物（中间可能经历了假体腔动物），生物复杂性增强，而这种增强，从现存的动物门中可以看出。



## 一棵新的动物演化树

不是每个动物学家都坚持上述观点，但几十年来这一直是一个流行的假说。一种最流行的理论将两侧对称动物分为两大类（原口动物和后口动物），对体腔的关注较少；不过，这一理论依然以“节”为依据，将节肢动物和环节动物归入关节动物。不过，在1988年，一条新的证据链被引入这个问题的研究中，由此很快发现，体腔发育假说和关节动物的概念，可能有很大的问题。美国印第安纳大学鲁道夫·拉夫领导的一个研究小组开始用基因序列数据来研究动物门之间的演化关系。基因突变会随着时间不断积累，所以物种之间DNA序列上的差异会反映出它们共享一个祖先的时长。亲缘关系较近的动物门的某一特定基因有相似的DNA序列，而关系较远的则有差异较大的DNA序列。拉夫及其同事把注意力集中在核糖体小亚基RNA的基因编码上，RNA是核糖体的组成部分，是所有细胞中都有的结构。这种基因的主要优势是，它存在于每一个动物物种中，做着同样的工作：帮助制造蛋白质。

1988年的研究，开启了使用DNA序列信息寻找真正的动物系统发生树的革新。尽管这项技术很新，分析方法还处于初级阶段，但有一个结论从一开始就很清楚。分节的环节动物和分节的节肢动物有着截然不同的RNA基因序列，并没有证据支持关节动物这一分类的存在。经过20多年的时间，来自更多物种的更多基因的DNA序列被确定，以计算机为基础的分析方法也得到了完善和改进。如今，最可靠的系统发生树包含了每种动物的100多个基因，它们显示出了非常一致的结果。这种“新的动物系统发生树”与过去的树之间有一些相似之处，但也有一些非常关键的不同之处。

在这个新的动物系统发生体系中，四个非两侧对称动物的门很早便从主干上分了出来，就如同在体腔发育假说和其他以形态学为基础的系统发生体系中一样。这意味着胚胎层数和对称性共同提供了准确的信息。水母、海葵、珊瑚、栉水母和海绵确实都是“基础”动物。在这些基础动物门分化之后，其他的动物成为两侧对称动物。两侧对称动物体现出新体系与过去的假说之间的区别。比如在新的系统发生树上没有仅仅由无体腔动物组成的区域，没有假体腔动物的位置，也没有仅仅由体腔动物组成的类别。相反，这三种类型的身体结构是混杂的。

20

由于这棵新树的很多不同部分都有体腔动物，这就意味着在动物演化过程中，体腔要么不止一次演化出来，要么消失不见，也有可能两种情况都存在。从实用的角度来看，这也许并不奇怪。充满液体的体腔为生活在许多环境中的无脊椎动物提供了优势——它们为身体提供支撑，同时像不可压缩的袋子，承受着不同肌肉的挤压。对于软体的动物来说，体腔增加了动物运动的力量和效率，使它们能挖掘、快速爬行，甚至游泳。从绘制演化树的角度来看，这意味着体腔不足以用于判断亲缘关系。分节也是同样。在某些环境中，身体分成几个单元会有好处，比如能提高运动效率，而且分节可能演化了不止一次。节就像体腔一样，在演化中很轻易地出现又消失，因而不能成为谁与谁更亲近的标志。并没有一个分类能被称作体腔动物，也没有一个分类能被称作关节动物。



那么，这棵由DNA序列构建的演化树是怎样的呢？在这个被迅速而广泛地接受的动物系统发生体系中，两侧对称动物被分成三大类，每一大类被称作一个“总门”。每个总门当中包含数个门。人类所属的总门名为后口动物总门。除了脊索动物门，后口动物总门中还包括棘皮动物门（海星和海胆）和半索动物门（臭烘烘的柱头虫）。在过去的演化体系中，也都有一个名为后口动物总门的大类，但基于DNA数据，其中包含的一些动物现在已经被移到了其他位置，最有代表性的便是毛颚动物门，或称箭虫。

两侧对称动物的另两大总门令人意外。以解剖学比较为基础，人们从来都没有猜测过它们的存在，它们也没有出现在过去 21 所有传统的演化树中。然而，DNA序列数据能有力支持这两个总门的存在。因为是近年才被提出的，所以这两类动物便需要新的名字。这两个名字都有点拗口。其中一类包括节肢动物门（昆虫、蜘蛛、螃蟹、蜈蚣）、线虫动物门（蛔虫）和其他数个门，被称作“蜕皮动物总门”。另一类包括环节动物门（蚯蚓、水蛭）、

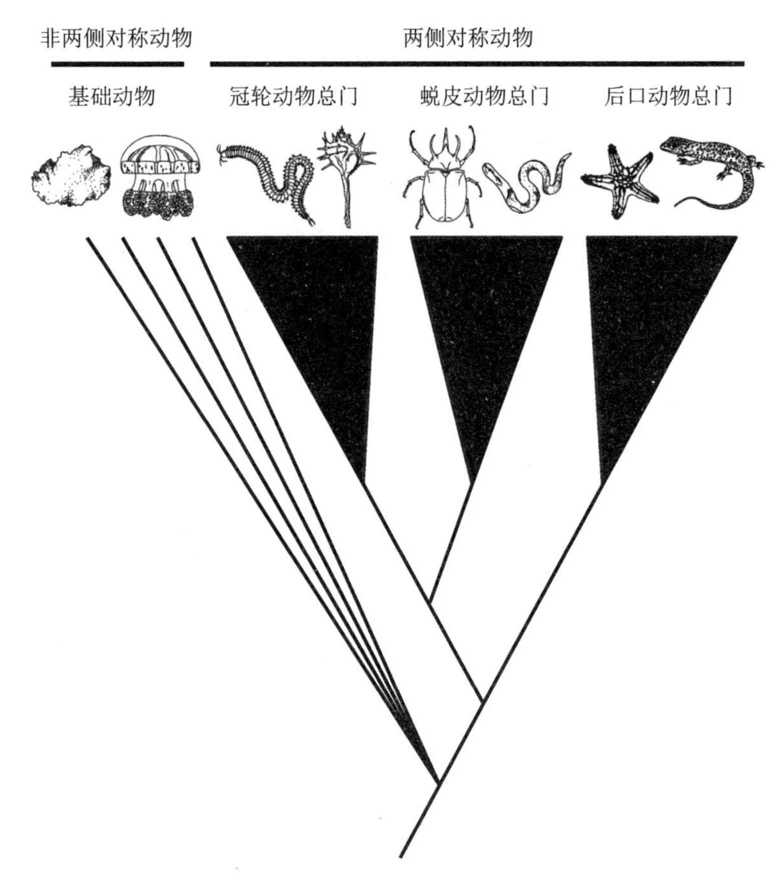


22 图2 基于DNA序列数据的新动物演化体系

软体动物门（蜗牛、章鱼）、扁形动物门（扁虫、吸虫、绦虫）、苔藓动物门（苔藓动物）和其他一些门，被冠以“冠轮动物总门”之名。

系统发生树最好的呈现方式是图示。如图2中所总结的，在新的动物系统发生树中，演化初期有四个非两侧对称的门分了出去，与两侧对称动物的大类分开。然后两侧对称动物又分成三个大的总门——后口动物总门、蜕皮动物总门和冠轮动物总门，图中均有呈现。顺便说一句，后两组在亲缘上最接近，它们合在一起，和过去一些树中的“原口动物”接近。有一点非常重要，在这三个大类中，并没有谁“高”谁“低”，因为所有的门如今都还存在，并没有不断上升的“自然阶梯”。在本书接下来的章节中，我们将审视每个分支上的动物，从非两侧对称动物的门开始，然后再依次研究两侧对称动物的三个总门。顺序是随机的。并不能因为人类位于后口动物总门中，就认为我们所处的类群在演化树上有任何特别的优先权。

第四章

# 基础动物：海绵、珊瑚和水母

海底完全都被遮住了，铺满了各种各样的珊瑚、海绵、海螺和其他海洋生物，体形庞然，千姿百态，色彩缤纷……这样的图景，连看数小时也不知疲倦，任何文字都无法描摹其非凡的美和魅力。

——阿尔弗雷德·拉塞尔·华莱士，

《马来群岛自然科学考察记》( 1869 )

## 多孔动物门：海绵

海绵是动物界所有成员中最不像动物的了。大多数海绵是花瓶状的，但有些看起来像是海洋里岩石表面不规则的块状增生物、湖泊和河流中的鹅卵石，或掉到水里的树枝。对海绵来说，严格的前后、上下或左右的概念是不适用的。它们没有明确的神经细胞和肌肉，但它们可以移动，非常缓慢地移动，而且——就像其他动物一样——海绵能对接触做出反应，并能察觉到环境中的化学变化。与其他动物不同的是，它们没有真正的口部或肠道，而是使用一个复杂的水流系统来捕获食物。要识别出海绵，可以以其表面的一个或多个大孔洞为标志，不过，它们的表面还有成千上万的稍小的孔。水流持续不断地从小孔流入，从大洞流出。水流中携带着溶解的氧气和细菌等食物颗粒，一种很重要的细胞参与了水流的形成，它们排列组成海绵体内由中空管道和空腔构成的网状结构。这些进食细胞被称作领细胞，各有一根能有节奏地运动的鞭毛，类似于前文提及的有鞭毛的领鞭虫。不过，它们的功能却不相同，因为——与领鞭虫不同——领细胞并不将领用作一张简单的网来捕捉食物。生有这些领细胞的腔室的横截面比孔的横截面大，这就意味着水流流入海绵后便会大大放慢速度。随着流入的水流几乎静止下来，海绵细胞便可以吞噬细菌和其他食物颗粒。

24

尽管海绵有不同类型的细胞，但大多数无法组成肾、肝、卵巢那样的有特定功能的器官（尽管领细胞的腔室可以被理解为简单的器官）。因此，海绵有时会被描述为具有“组织级别”的身体结构。有些海绵有惊人的再生力量，令人瞠目结舌，科幻电视剧集《神秘博士》中能再生的外星人就是以海绵为灵感的。揭示这一特性的决定性实验过程是由美国北卡罗来纳大学的亨利·范·彼得斯·威尔逊于1907年发表的。威尔逊将一个活体海绵捣碎，然后用一块筛面粉的细布筛了一遍，由此将海绵的主体分割成了独立细胞。之后，威尔逊观察到，这些细胞缓缓地移动，重新聚集在一起，形成了一个新的海绵！如果两个不同品种的海绵的细胞混杂在一起，它们会自行分开，再生为两个原始的海绵。尽管动物界的许多分支都有再生能力，但其他动物都无法和海绵相提并论。

海绵的外层细胞和内层细胞之间有一层“结缔组织”，还有名为海绵硬蛋白的坚韧纤维或由碳酸钙、二氧化硅形成的矛状物或星状物（即骨针）加强其连接功能。有海绵硬蛋白框架但没有骨针的海绵，曾被广泛用于洗涤和清洁，不过现在大部分都已被合成泡沫取代。角骨海绵属和马海绵属均为此类。海绵的采集和使用可以追溯到很久之前。公元1世纪，老普林尼很详细地描述了如何使用海绵来清洁伤口、减轻肿胀、止血和治疗叮咬。更早之前，公元前4世纪，亚里士多德便描述了头盔的内衬应该使用何种海绵：

> 阿喀琉斯海绵格外细腻，质地紧密而结实。这种海绵被用于给头盔和护胫甲做内衬，以减弱击打时的声响。



值得注意的是，将海绵作为工具使用的并非只有人类。在澳大利亚西海岸的沙克湾，一种宽吻海豚学会了从活体海绵上撕扯下碎片，附着在自己的鼻子上，以便在海底沙滩上觅食时获得保护。

海绵构成了一个门：多孔动物门。门下分为三个纲：寻常海绵纲（包括浴海绵）、钙质海绵纲（拥有碳酸钙针状物）和稀有的见于深海的六放海绵纲。六放海绵又名玻璃海绵，外形格外漂亮，并与其他种类的海绵有一些重要区别。其中一个特征便是它们的身体大部分都是由“合胞体”构成的，合胞体是一层含有许多细胞核而细胞质未被细胞膜分隔为独立细胞的结构。同样不寻常的是，它们的二氧化硅骨针被编织成精致的格栅状结构，就像精美的玻璃笼子。其中最广为人知的代表是阿氏偕老同穴，俗称“维纳斯的花篮”，它在太平洋海底的岩石上附着生长，拥有一个约30厘米长的圆筒形塔状骨架，由错综复杂的硅质骨针构成。通常，会有一对活虾被困在硅质的笼子中，一只雄性，一只雌性，因为长得太大而无法从海绵的骨架中游出去。不过，这对虾的后代能够从网格中游出去，游到其他的阿氏偕老同穴的内部生活，只留下它们的父母偕老于此。日本有一个古老的习俗，人们将这种海绵的标本作为结婚礼物，它象征至死不渝的结合。

26

## 古怪的丝盘虫

海绵并非唯一一种没有三条明确对称轴的动物——三条对称轴是指头尾对称轴、上下对称轴和左右对称轴（即两侧对称轴）。还有三个动物门的身体结构也是“非两侧对称”的：刺胞动物门（海葵、珊瑚、水母）、栉水母动物门（栉水母）和扁盘动物门。最初只有一个物种被归入扁盘动物门中，那是一种小小的煎饼状的生物，名为丝盘虫，拉丁文学名 *Trichoplax adhaerens* 的意思即为黏黏的、多毛的盘子。不过，近年的基因分析表明，它并不是孤家寡人，实际上，还有很多与这种小生物近似的物种，浮游在从太平洋到加勒比海以及从地中海到红海的热带和亚热带海域中。乍看之下，丝盘虫很容易被误认为大号的阿米巴原虫，它的身体直径在半毫米到一毫米之间，但如果仔细查看，就能看出它由数千个细胞组成，足以成为真正的动物。它呈不规则的扁平形状，并没有特别明显的前后端之分。它通过改变身体形状和摆动覆盖身体下部的数以千计的微小纤毛，可以在坚硬的物体表面向任何一个方向爬动。丝盘虫没有口部和肠道，身体底部的表皮能分泌酶，酶将单细胞藻类等食物分解为营养物质，然后吸收。简言之，扁盘动物是非常不同寻常的动物，长久以来都令动物学家十分迷惑。

扁盘动物最初由德国动物学家、海绵专家弗朗茨·艾尔哈德·舒尔策于1883年发现，但有意思的是，他并不是在大自然

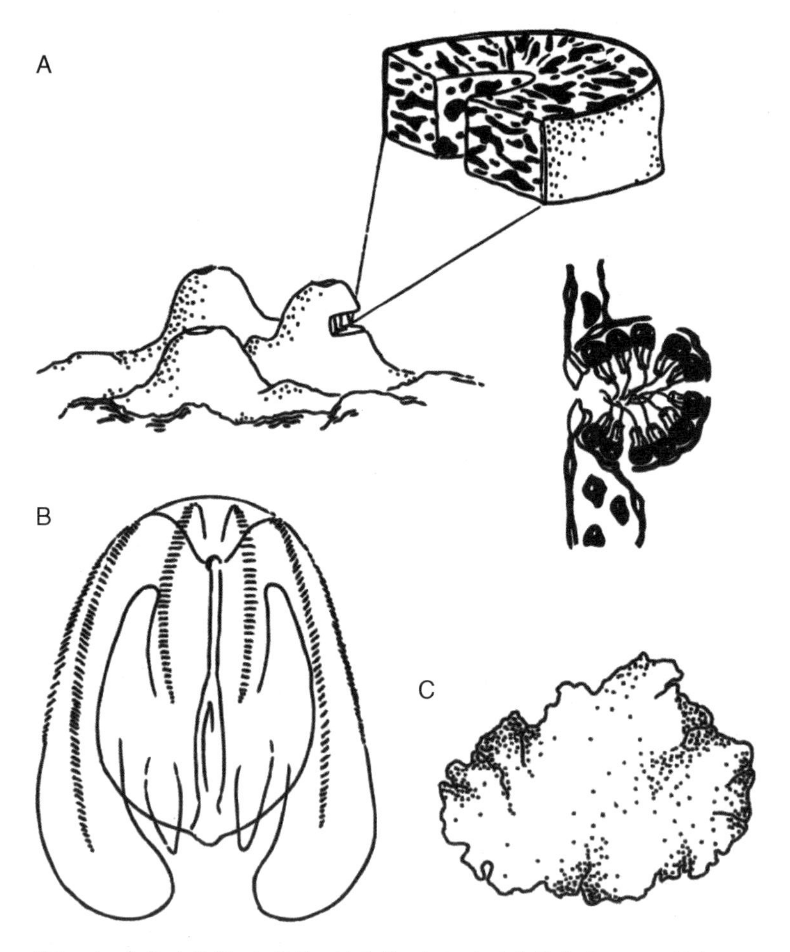


图3 A. 多孔动物门，即海绵：蜂海绵，领细胞腔体结构展示；B. 栉水母动物门：淡海栉水母；C. 扁盘动物门：丝盘虫

中发现丝盘虫的。他在奥地利一家海洋水族馆的玻璃壁上发现了这种新物种在爬行，这便意味着，起初人们并不知道它在野外 27 生活在什么地方。事实上，很多动物学家后来都声称舒尔策认为丝盘虫是一种新物种完全是搞错了，坚持说它不过是某种众所周知的海葵类动物的幼虫。要到几乎100年之后，舒尔策的正确性才得到完全证明，对野生的和实验室中的扁盘动物的广泛研究已经证明它们足以构成一个独立的门，尽管物种数量不多。28 不过，那些想仿效舒尔策的人要小心了。我有一次在一个水族商店里拿放大镜观察鱼缸里的浮渣，结果却被恼羞成怒的经理赶出门去。



## 栉水母门：栉水母

栉水母门是第三个非两侧对称的动物门，在身体构造上与海绵或扁盘动物存在很大差异。栉水母是一种捕食者，它在海中缓缓移动，吃掉其他缓缓移动的动物，如其他栉水母、甲壳纲动物以及海洋生物的幼虫。和大多数捕食者不同，栉水母不会追逐或跟踪猎物。它们只是在移动中碰到小型浮游生物，然后用特化细胞分泌的小滴黏胶捕获它们，黏胶通常集中在口部两侧的两条长触须上。与海绵和扁盘动物不同，栉水母门动物拥有神经细胞和平衡感器官，能够迅速对环境做出相应的反应。尽管大多数栉水母不过是几厘米大小的果冻状的一团，但几乎所有见过活着的栉水母的人都会将其列为地球上最美丽的动物之一。它们最明显的特征便是身体两侧呈带状排列的八个“梳栉”，每个上面都有数以千计的纤毛。这些纤毛以高度协调的方式摆动，每一根都紧随着邻近的纤毛移动，形成一种“节奏波”，十分像是足球比赛间歇时偶尔会在球场周围出现的“墨西哥人浪”。成千上万根微小纤毛轻柔漂动，推动着栉水母缓慢而悄无声息地在海洋中穿行，在这个过程中会散射光线，创造出不断变化和闪烁的彩虹。栉水母中最广为人知的是葡萄大小的球栉水母，俗称“海醋栗”，在太平洋和大西洋中以及英国海岸上均有发现的侧腕水母便属此类。但最特别的栉水母无疑该是体长达一米的巨型的带栉水母，这种水母又被称作“维纳斯的腰带”或“爱神带”，得名自罗马神话中的爱神。这种栉水母不是通常的卵状，而是有着细长的带状身体，当光线被它身体上一排排的纤毛散射开来，带状的身体会呈现出令人目眩的彩虹色，在海水中闪闪发光。用理查德·道金斯的话来说，带栉水母“女神都不配佩戴”。



除了在海洋食物网中扮演一个不重要的角色外，大多数栉水母对人类几乎没有直接影响。不过，有一个物种是基础无脊椎动物中的异类，一个大反派。20世纪80年代，大西洋的栉水母淡海栉水母偶然被引入黑海，可能是随着商业海运的压舱水进入的。一到新环境，远离了自然竞争者和天敌，它就开始迅速繁殖——同时还消耗掉了大量的幼鱼和甲壳动物幼体。一些（有争议的）估计认为，黑海中这种微型栉水母的总重量超过5亿吨。当地本来已经承受着沉重捕捞压力的鳀鱼渔业出现大幅减产。生态学家们为了如何解决问题争执不休，这个过程中，出现了一个完全意外的解决方案：又一次偶然引入了一个新物种。这次新来的物种是另一种栉水母，名为瓜水母，这种水母十分贪吃。碰巧的是，瓜水母不吃鱼或甲壳动物，而是专门捕食其他的栉水母，别的物种都不吃。随着入侵的瓜水母尽情享受淡海栉水母盛宴，鱼类数量出现了缓慢但可喜的恢复迹象。

## 刺胞动物门：刺和超个体

在四个“非两侧对称”动物门中，海绵和扁盘动物都不具备精准的对称，而栉水母具有“两侧辐射”对称，这意味着它们的身体通过180度旋转对称。第四个非两侧对称动物门也是最大的一个门，名为刺胞动物门，包括水母、海葵和珊瑚等一些非常为人熟知的动物。这些动物的身体同样不具备头尾对称轴、上下对称轴和左右对称轴，不过同样符合辐射对称或旋转对称，只有少数例外。刺胞动物身体的基本构造是杯状或瓶状的，身体的一端有一个大开口，这个口既充当口，也充当肛门。开口周围是一圈触手，每一根上面都武装着数以千计的带刺的细胞，这种细胞被称作“刺细胞”。这些细胞在被接触后三毫秒内会发射出带倒刺的小叉子，也就是刺丝囊。刺丝囊通常是带有毒素的，是刺胞动物用于攻击和防御的主要武器。

30

刺胞动物有神经细胞，而且就像栉水母一样，这些细胞围绕着身体排列成格栅状网络，而不是像大多数其他动物那样形成一个单独的脑部和“中枢神经索”。形成身体的细胞层包括在外部的外胚层和在内部的内胚层，中间隔着一种名为中胶层的物质。中胶层基本上是由蛋白质构成的，而非活细胞层，但许多刺胞动物的中胶层中依然有零星的可以移动的细胞，某些物种的中胶层中甚至有肌肉细胞形成的可以伸缩的纤维结构。不过，中胶层中的细胞并没有形成复杂的器官，所以通常认为刺胞动物的身体仅由两个基本细胞层构成。

刺胞动物分为四类。第一类为珊瑚纲，其中包括海葵，比如可以在海滩岩石形成的潮水潭中发现的颜色鲜艳的沟迎风海葵和等指海葵。这些海葵的身体只有一个朝上的开口，另一面身体黏附在岩石上，但十分不牢靠。一旦被涌入的潮水淹没，海葵就会张开顶部的触手冠，等待微小的猎物漂到或游到它们附近，然后迅速将其蜇咬并吃掉。虽然通常是静止的，但海葵并不是完全固定的，它们可以离开所在位置，通过漂流或缓慢的游动转移到其他地点。它们还能够利用那个起附着作用的单足缓缓移动，有时是为了寻找一个更适合摄食的位置，有时是因为与其他海葵发生激烈凶猛但动作缓慢的战斗，两只海葵用武装了刺丝囊的膨胀的“棒子”相互蜇刺。

31

珊瑚也属于珊瑚纲，它们展示出了在动物演化中反复出现的一个特征：集群性。珊瑚是由成千上万甚至上百万的微小动物组成的，每一个个体都像是一只直径只有几毫米的小海葵，但它们相互连接，形成一个巨大的“超个体”。活珊瑚的生长，是由于微小的“个虫”出芽繁殖，因此整个集群有着相同的基因构造。它是一个巨大的克隆体。有些珊瑚物种的集群呈扇状散开，有些如鹿角一般分支，还有些长得像是恐怖的身体器官，比如片脑纹珊瑚和别名为“死人手指”的指状海鸡冠珊瑚。不过在所有珊瑚中，最引人注目的是造礁珊瑚，出芽的个虫身体周围会分泌出碳酸钙，形成巨大的白垩结构，许多其他动物都将其作为自己的家园。

刺胞动物的第二类为水螅纲，其外形与海葵十分相似。水螅纲中包括一些大型的颜色鲜艳的海洋物种，另外还有在池塘和河流中的小型水螅。尽管水螅的英文名hydra来自希腊神话中一个长了很多头的水怪的名字，但它的身体是一根只有几毫米长的小管子，有一个开口向上的口部，口部周围有带刺的触手。所有品种的水螅都能捕捉、食用小型的淡水无脊椎动物，但有些品种还具有其他技能。绿水螅奴役了一种单细胞藻类，这种藻类生长在它的肠细胞中，令其身体呈现出明亮的绿色，并通过光合作用为水螅提供食物。一些水螅也如同某些珊瑚纲动物那样，以彼此连接的集群形式生活。名声不佳的僧帽水母（别名“葡萄牙战舰”）就是由巨大的水螅集群构成的充气的漂浮体，其下有数以千计的彼此相连的个虫，形成10米长左右的一串，布满有毒的刺丝囊，充满威胁地摇摆不定。

刺胞动物如果长着向上的口部，如成年海葵、珊瑚和水螅那样，便是“水螅体”。若是与此相反，开口朝下，则被称作“水母体”，这是钵水母纲典型的身体构造。很多刺胞动物的生命周期是在口朝上和口朝下两个方向间转换的。水母体和水螅体除了单纯的口器方向不同外，还有其他一些不同。近年来利用基因表达模式的研究显示，向下的水母触手和向上的水螅触手的结构实际上是不同的。水母和所有的刺胞动物一样，都是捕食者。这种铃铛形状的胶质动物以身体外壁有节奏的脉动为推力，在海洋的表层漂浮或缓缓游动。海洋表层的水中有大量浮游生物，如甲壳动物和鱼类幼苗，水母摆动长着有毒刺丝囊的触手来捕捉这些生物。很多游泳者都有过与水母的触须意外擦身而过后生出痛苦皮疹的经历。水母这一基本类型演化出很多种变体，其中最不寻常的一种是根口水母。这种水母并没有一个向下的口部，其口部已被融合的组织封闭起来，取而代之的是八条分支腕上许多像口部的小开口，每一个开口都通过复杂的管道系统连接着肠道。很多根口水母目动物，如巴布亚硝水母，通过身体组织中含有的数百万能够通过光合作用产生能量的共生藻类来补充食物摄入。硝水母的密度因而能高到令人难以置信的程度。在太平洋帕劳群岛的埃尔马尔克岛上有一个“水母湖”，巴布亚硝水母在湖中密集生长，有时能够达到每立方米的海水中包含1 000只这种六厘米长的动物。

对人类来说，比真正的水母，甚至“葡萄牙战舰”更加危险的，是刺胞动物门的第四类动物：立方水母纲。由于身体的形状，它们也被称作箱形水母，在热带沿海地区最为常见。与真正的水母不同的是，每个立方水母有24只眼睛，其中六只具有晶状体、虹膜和视网膜，能够形成远处的物体的图像。有些物种，33比如有“海黄蜂”之称的澳大利亚箱形水母，拥有极为强效的毒液，足以令游泳者心生畏怯。“海黄蜂”的刺有剧毒，对人类来说

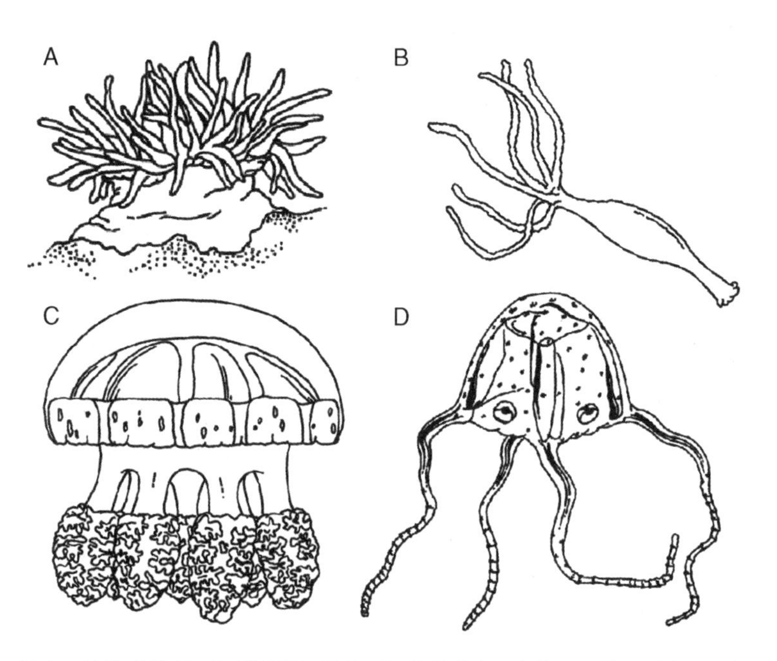


图4　刺胞动物门：A. 珊瑚纲，海葵；B. 水螅虫纲，水螅；C. 钵水母纲，巴布亚硝水母；D. 立方水母纲，即箱形水母，或称伊鲁坎吉水母

甚至也是致命的。其他一些立方水母的刺在接触时不会引发那么大的痛苦，但能引起一种不同寻常的延迟反应，这种反应被称作“伊鲁坎吉综合征”，这是根据昆士兰北部海岸澳大利亚土著人的名字命名的，那里箱形水母十分常见。被伊鲁坎吉箱形水母蜇伤的游泳者会逐渐出现剧烈的背痛、肌肉痉挛、恶心、血压升高以及一系列令人不安的心理效应，会产生“一种末日即将来临的感觉”。

34

第五章

# 两侧对称动物：构筑一个身体

人类不过是蠕虫。

——爱德华·林利·桑伯恩，

《笨拙》(1881)

## 有前端的生命体

你是两侧对称动物，鱼、鸟、蠕虫、鱿鱼、蟑螂和其他千千万万种动物也都是。实际上，大多数动物都是两侧对称动物。两侧对称动物门是动物界的大分支，顾名思义，这些动物都有一条在身体中心延伸的镜像对称轴。这条对称轴将身体分为左侧和右侧，这就意味着身体必须有明确的前端和后端，顶部和底部，而前后、顶底是不对称的。就人类而言，左侧右侧如你所见，而你身体的前端（前部）实际上是指你的头，后端（尾部）是指你坐的部位，你的顶部即背部，是沿着你的脊椎的部分，而底部即腹部，是你的肚子。这些方向其实很容易理解，请记住，从演化的角度看，我们人类开始直立行走并不是太久的事情。

35

两侧对称与在大多数刺胞动物、栉水母中发现的旋转对称不同，也与扁盘动物和海绵缺乏清晰的对称结构不同。两侧对称动物身体组织中的对称性，并不只表现在皮肤表面。两侧对称动物有明确的肌肉块，可用于主动运动，几乎所有的动物都有集中的神经索和位于前部的脑，以及集中在前端的专门的感觉器官。大多数都有一个管状的或“贯通”的肠道，口部和肛门分开，可以有效地处理食物。只有少数例外，这些物种只有肠道有一个开口，但可能会发生二次演化，再次演化到前述状态。两侧对称动物的演化，标志着拥有主动的、有力的、直接的运动能力的动物的崛起，它们以一系列的感觉器官去面对前方的环境，可以挖掘、爬动、游动，同时将排泄废物甩在身后。两侧对称动物可以真正地从三个维度探索和开发这个世界。

两侧对称动物即三胚层动物，它们与更“基础”的动物门之间的区别，早在一个世纪前就已经被注意到了。1877年，知名的英国动物学家雷·兰基斯特将两侧对称动物的胚胎和刺胞动物、海绵等的胚胎进行了对比并指出，在早期的发育过程中，两侧对称动物多出一层细胞，而这层细胞将发育成成体动物的轮廓分明的肌肉块。胚胎内部和身体对称性上的相似性当然是最基本的。到了20世纪末，生物学家们吃惊地发现，这种相似性的深刻远超过去的想象——一直深入到DNA。研究发现，所有两侧对称动物都使用同一组基因来构筑身体，这个发现是20世纪最有意思的科学突破，改变了20世纪80年代以后的生物科学。这是一个具有爆炸性影响的发现，不过这场爆炸性革命的导火索燃烧得并不快。

## 异形发育与同源基因（Hox基因）

被人们铭记至今的威廉·贝特森是基因科学的奠基人之一。刚获得学士学位，距离成名还有很久的时候，年轻的贝特森发表了一系列关于柱头虫解剖结构的科学论文，柱头虫是一种海洋无脊椎动物，它在演化中的位置当时还没有确定。贝特森的研究赢得了一些赞誉，不过他并未就此满足，他说这对揭示出演化到底如何运作几乎没有启发。在一封给母亲的信中，他写道：

36

> 五年之后，不会有任何人觉得那项研究有什么意义，而这种轻视是非常正常的。它与我们想要了解的事情没有什么联系。我只是在幸运的时候碰上了这项研究，又恰好在人们需求最高时兜售出去了。



贝特森真正想要了解的，是物种之间的变异到底是如何形成的。于是，接下来的八年中，贝特森全身心地投入了为动物和植物的“变种”进行编目的工作，并于1894年出版了巨著《变异学研究资料》。这部书值得珍视的成果众多，其中贝特森讨论了一种不同寻常的变异，在这种变异中，某种动物身上的某个“结构”被通常该出现在身体其他部位的结构替代，如在该长眼睛的地方长出一根触角。这种奇怪的“同源异形”的变异现象一直未能引起广泛关注，直到1915年，卡尔文·布里奇斯指出果蝇身上的这种变化会传递到后代身上。遗传是关键。它指向了基因。这就意味着必然有确保身体器官正确发育的基因，当这些

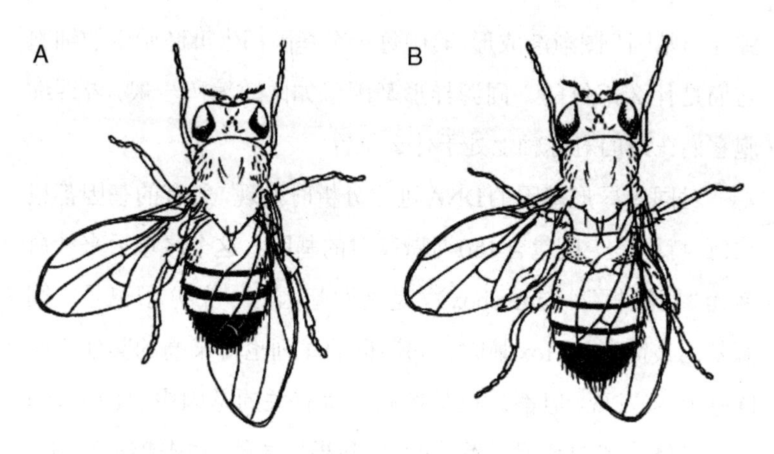


图5　A. 正常果蝇，B. 卡尔文·布里奇斯的“双胸”突变果蝇

基因中的某一个发生了错误—— 一次突变时，指令就会被误读。身体的一个区域会按照另一个区域的方式发育。在最初的同源异形突变中，翅膀，或者确切地说，一部分翅膀长在了它们不该长的位置。布里奇斯将这种突变称作“双胸”。双胸这样的反常实在太大，以至于难以在演化过程中发挥任何直接作用，在解剖结构上发生如此巨大变化的蝇是无法在自然环境中生存下来的。但是突变基因为如何构筑身体提供了线索，而这自然有助于理解动物演化。

另一位基因学家埃德·刘易斯继续研究这一发现，投入了极大的精力和耐心。刘易斯发表了一系列精彩的论文，其中包括后来为他赢得诺贝尔奖的发表于1978年的杰作。他通过这一系列论文指出双胸变异并非绝无仅有。好几个基因可以导致同源异形变异，每一个突变都会影响果蝇身体的不同区域，并且都对应果蝇某条染色体的某一部分。另一位基因学家汤姆·考夫曼发现了控制身体头部和前端发育的基因，因而一幅完整的“同源异形基因”图渐渐成形，其中每一个基因都告诉胚胎中的细胞它们是什么部位的。同源异形基因就如邮政编码一般，告诉细胞它们在蝇的对称轴上处于什么位置。

对同源异形基因的DNA进行分析时发现，所有的基因都很相似，特别是一段包含180对碱基对的基因。这个区域后来被命名为“同源框”（homeobox），是同源基因的一个分子记号。同源基因被简称为Hox基因。同源框的序列也在果蝇的其他一些DNA中被发现，但都是在与控制发育有关的基因中，比如参与果蝇身体分节的基因。不过很快，除果蝇之外，越来越多的物种被纳入研究。即将揭开的奥秘会带来怎样的冲击，生物学家几乎都没有预料到。1984年，瑞士的巴塞尔成为焦点，比尔·麦金尼斯、迈克·莱文、黑岩笃志、厄恩斯特·哈芬、里克·加伯、埃迪·德罗贝蒂斯、安德烈斯·卡拉斯科、瓦尔特·格林等研究人员组成的一个充满活力的组织，打破了生物学知识的界限。麦金尼斯和同事们测试了能否在其他动物身上提取的DNA中检测到同源框序列，所取得的成果令人震惊。不仅其他昆虫有同源框，最初的实验表明，蠕虫、蜗牛、老鼠，甚至人类可能也有。卡拉斯科、麦金尼斯、格林和德罗贝蒂斯很快将一只青蛙的同源基因分离并测序，他们证明了这一点：脊椎动物确实有同源基因。

科学界一片热议，激动不已。顶级期刊争相展示每一个新发现，热切的读者如饥似渴地阅读每一篇公开发表的论文。每一次会议和研讨会的主题都是同源框。在伦敦的一次科学讲座上，一位科学界同人报告说发现了一种新的发育控制基因，我记得当时人们提出的第一个问题是：“它有没有……我要不要说出那个神奇的词？”我甚至认识数位科学家就此放弃了自己终生研究的方向，转而投入了同源基因的研究。

发现同源框，以及发现同源基因广泛存在于苍蝇、青蛙等各种各样的动物中，这开启了一项革新。1984年之前，人们基本上不了解大多数物种的基因如何控制形态发育。同源基因可能提供一个解决这个问题的新途径吗？这个问题促使乔纳森·斯莱克将同源基因的发现比作1799年埃及罗塞塔石碑的出土，罗塞塔石碑为古代书写系统提供了最早的翻译。我们现在是否有了一个方法来比较迥异的物种对胚胎发育的控制？这种乐观想法并非人人认同，但事实证明这一想法有很好的根基。青蛙和人类等脊椎动物的很多同源基因，实际上相当于果蝇的同源异形基因，也就是Hox基因，从根本上说，它们发挥的作用是相同的。如同在果蝇的身体中那样，我们自身的Hox基因也如邮政编码一样，告诉人类细胞它们该处于头尾轴上的什么位置。



对于演化生物学来说，其意义是巨大的。如果脊椎动物和昆虫都有Hox基因，那么毫无疑问所有的两侧对称动物都有Hox基因吧？如果脊椎动物和昆虫都用这些基因来表明它们在身体主轴上的位置，那么这个属性也必须追溯到两侧对称动物的起源。我们可以对这些观点充满信心，原因很简单，蝇和人类的共同祖先同样也是所有蜕皮动物、冠轮动物和后口动物的祖先。无腔动物门可能是两侧对称动物里较早分化出去的，但最近的证据表明，即便是在这些动物体内，Hox基因也发挥着相同的作用。因此，动物界的这一整个分区，有着分明的前端和后端的29个动物门，三维世界的主动探索者，都在使用相同的基因组来构建从头到尾的主轴。

## 上与下，左与右

那么身体的另外两条轴，上下方向和左右方向的情形又是如何呢？同样，确保细胞了解自己在这两个方向上的位置的基因也被发现了。而且结果证明，就如同Hox基因的情形一样，存在巨大差别的两侧对称动物使用的是本质上相同的基因——只有一处有趣的变化。在蝇类的胚胎中，底部也就是腹部的细胞会形成主要的神经索，sog基因在这个过程中起着关键作用。而上部也就是背部的细胞会形成表皮细胞，这个截然不同的命运由dpp基因控制。脊椎动物同样有sog基因和dpp基因，只是名称不同。脊椎动物的sog基因被称作脊索蛋白，在要发育为背部的一侧表达，也就是我们的神经索所在的位置；BMP4是脊椎动物的dpp基因之一，在腹部表达，就方向而言，正好和蝇类相反。在更多的物种身上比较这些基因，结果发现大多数动物的方向定位都和蝇类一样；而我们人类所属的动物门即脊索动物门，则是颠倒的。对于左右方向轴的演化，我们目前了解得较少，但我们已经知道至少有两种基因，分别是nodal基因和Pitx基因，参与了蜗牛和人类这样迥异的动物中这条轴的构建。



40

这种相似性并非仅限于身体的方向，而是深入了内部。比如脊椎动物身上数个控制心脏构造的基因在昆虫身上也有发现，这些基因在昆虫身上也控制了一个脉动的肌肉管的发育。一些基因系统指明了眼睛将在哪里形成，无论苍蝇、蠕虫还是人类，这些基因中的大多数都是相同的。这些惊人的发现似乎共同表明了所有两侧对称动物的早已灭绝的祖先都有一个基因系统来区分背部和腹部，辨识左侧与右侧，指示细胞在头尾方向轴上的位置，以及参与构建各种内部的结构和感觉器官。这些基因及其作用已经保留了数亿年，不过经历了一些修改，并且我们的一个祖先由于某种原因经历了上下颠倒。法国的博物学家艾蒂安·若弗鲁瓦·圣伊莱尔在1830年就曾经提出过这些观点，只是他的观点基于掺杂了想象的解剖学比较和有些可疑的理想化基础。若弗鲁瓦说："从哲学上来讲，仅有一种动物。"他活着时，他的观点始终未得到认可，然而至少在两侧对称动物方面，他的观点是正确的。

这组用于构筑身体的古老基因组有时也被称作"发育工具包"。工具包中的一些基因，如Pitx基因和Hox基因，为与DNA结合的蛋白质编码，负责打开或关闭其他组的基因。其他一些基因则为在细胞间传递信号的分泌蛋白编码（如nodal基因和dpp基因）或干预信号（如sog基因）。然而，这些仅仅是冰山一角，工具包中包含成百上千个为与DNA结合的蛋白质编码的基因，数十个促生分泌因子的基因，此外还有负责为分泌因子所结合的受体编码的基因。所有这些基因在相差极大的两侧对称动物中都可以找到，尽管有时个别基因在特定动物族群的演化中消失了。如前面的例子所示，不同门的基因的作用通常是相似的，不过在某些情况下，工具包基因在不同的族群中发挥的作用是不同的。这些便是构建两侧对称动物的古老基因。但它们是怎么产生的？发育工具包的演化能否帮助我们了解动物演化的初期阶段发生了什么呢？

深入研究非两侧对称动物——海绵、扁盘动物、刺胞动物和栉水母的基因组序列，会得到更丰富的信息。有些关键基因可以在所有动物身上发现，但很多则不能。刺胞动物可能是非两侧对称动物中最接近两侧对称动物的，它们拥有几乎全部的工具包基因，不过它们的Hox基因簇没有那么复杂。其他的门缺少的工具包基因要多一些，比如海绵完全没有Hox基因。走出动物界，研究一下鞭毛虫，我们会看到更大的差异：很多的工具包基因缺失。结论是一目了然的。构建动物身体所必需的基础基因组是在多细胞分化发生时演化出来的，在动物演化的初期，这组基因得到了扩展，变得复杂起来。五亿年前，在两侧对称动物时代的黎明时刻，一个庞大的发育基因工具包就已经存在。今天，用这些基因来塑造身体的动物难以计数，包括两侧对称动物的所有三个大类：冠轮动物总门、蜕皮动物总门和后口动物总门。

第六章

# 冠轮动物总门：奇妙的蠕虫

人们可能会问，是否还有许多其他动物像这些低等构造的动物一样，在世界历史中扮演了如此重要的角色。

——查尔斯·达尔文，

《腐殖土的产生与蚯蚓的作用以及对其习性的观察》(1881)

## 环节动物：有生命的耕犁和吸血者

在查尔斯·达尔文去世前一年，他出版了他的最后一本书。这部作品备受欢迎，至少最开始的时候是，其销售速度甚至比《物种起源》还快。这部不像畅销书的畅销书《腐殖土的产生与蚯蚓的作用以及对其习性的观察》，包含了达尔文根据自己断断续续进行了40多年的实际研究得出的见解。其时，他已经是一个祖父，明白自己垂垂老矣，很渴望能在"与蚯蚓为伴之前"出版自己关于蚯蚓的发现。书中最重要的结论是蚯蚓不应被视作在修剪整齐的维多利亚时代草坪上留下难看粪便的害虫，事实上，它们是"有生命的耕犁"，对土壤健康至关重要。达尔文指出，蚯蚓将树叶等有机物质拖到地下，它们造成的孔道能给土壤通气，提供排水的渠道，它们的行动能促进土壤混合，防止土层板结，从而促进植物的生长。蚯蚓能磨耗岩石和石头，从而对地质产生影响。它们还能掩埋古代的遗迹，对考古也有影响。

43

蚯蚓属于环节动物门，它们的解剖结构使它们能产生这样的影响。环节动物拥有柔软的身体，肌肉发达，身体很长，一端是口部，另一端是肛门。它们的身体中有很多充满液体的空间，即体腔，内部的水压使它们有一定的硬度，同时又极具柔韧性。所有这些特征都有助于它们在泥土中穿行、钻洞，甚至通过肠道来移动泥土。不过，在这个过程中，起到关键作用的环节动物的一个特征，无疑便是身体分为一系列单元，也就是环。这是一个显著的特征，一眼就能看到，令环节动物名副其实。身体分节，每一节都有自己的肌肉、体腔和神经控制系统，令蚯蚓能在收缩部分身体使其变得又长又细的同时，挤压身体的另一部分使其变得又短又胖。细的部分向前探入缝隙，而粗的部分固定身体，然后从头到尾传递收缩的波动，身体向前推进，穿过泥土。



环节动物门的蠕虫数量超过15 000种，大多数生活在海洋或淡水中，而非陆地上。在它们的演化多样化进程中，分节是关键。比如掠食性的海生沙蚕可以在收缩左侧一部分节的同时收缩右侧的一部分节，将身体扭曲，以形成左右摇晃的波动。这种迅速而协调的扭动能推动沙蚕快速向前，使其可以追逐并捕捉到猎物。其他海洋环节动物则要被动一些，它们生活在洞穴或孔道中，不爱活动，过滤海水中的微粒为食。不过，即便是这样的虫子，也能高效地驱动分节，它们用协调的收缩波来冲洗洞穴或孔道，引入新鲜的富含氧气的水。44

不过，一类很有名的环节动物失去了所有的分节，而这也是有充分原因的。这类动物便是蛭纲动物，也就是水蛭。有些水蛭是捕食者，吞食小型的水生无脊椎动物。另外一些如每个热带探险者或影迷所知，会附在大型动物的肉上吸食它们的血液。这些寄生水蛭使用强有力的吸盘附着在皮肤上，射出一种强效的抗凝血剂防止血液凝固，用三瓣刀刃一般的颚划开血肉。马、鹿、鱼和人腿这样的食物来源并不常有，如果有机会碰到，水蛭通常会饱餐一顿，而这样就影响了分节。水蛭从类似蚯蚓的水生环节动物演化而来，但已经失去了身体内部的环节分隔，也就是“隔膜”。因此水蛭可以在吸食血液时伸展身体，像气球一样膨胀起来。这种改变有其缺点，导致水蛭无法像蚯蚓或沙蚕那样协调地收缩波动，其中大多数物种需要依靠效率低很多的环形移动方式。



某些水蛭吸食人类血液的能力，尤其是其不会引发痛苦的能力，已经被医学利用了很多个世纪。2000年前，希腊医生老底嘉城的塞米森曾经记载过使用水蛭来放血的事情，这项操作在世界许多地方一直沿用到19世纪。水蛭可用于释放“坏血”，纠正“不平衡的体液”，而“不平衡的体液”被视作很多病因不明的疾病的源头。水蛭的英文单词leech源于leoce，在盎格鲁-撒克逊语中意为“医生”或“治疗”。18世纪和19世纪，水蛭的需求量巨大，大型的医用水蛭品种医蛭的自然种群被过度利用，以至于直到今天，在大部分欧洲地区，这个物种依然非常稀少。水蛭养殖场蓬勃发展，但即便如此，依然满足不了需求。这其实并不奇怪，因为到19世纪30年代，法国每年要进口4 000万只水45

蛭，数量实在惊人。但水蛭并非只出现在医学史中，近年来对它们的应用又出现了神奇的回潮。现代显微外科手术中常见的并发症是静脉功能不全，当外科医生能够修复动脉但不能修复细小且血管壁很薄的静脉时便可能发生这种情况，会导致在组织重建或重接过程中血压升高。这时可以使用水蛭吸食掉多余的血液并射出抗凝血剂来缓解，为组织的自然愈合提供充分时间。这种技术已经在重接或修复眼睑、耳朵、阴茎、手指和脚趾的手术过程中成功应用。

如今被分在环节动物门的，还有三大类蠕虫，每一类过去都独立成门，直到通过分子分析确认它们的演化位置。这三类是螠虫（又称勺虫）、星虫（又名花生虫）和生活在深海的须腕动物（又名胡须虫）。前两类并没有分节，它们很可能是和水蛭一样，在演化过程中失去了这一特征，只是程度更加极端。曾经有很长一段时间，须腕动物也被认为是不分节的，直到1964年，某些从深海海床上采集的样本被发现拥有分节的短尾巴，这一部分被称为“后体部”。于是人们突然间意识到，之前所有的描述都——相当尴尬地——基于不完整的样本。很多须腕动物生活在泥泞的洞穴中，体形极瘦，身体延展开也不过几厘米长。不过，还有一些须腕动物体形硕大，构造出竖直突出的管道附着在海洋深处的岩石上。

1969年，美国海军在下加利福尼亚海岸附近进行深海潜水器操作时首次发现了大型管状蠕虫。这些动物被命名为巴勒姆氏瓣臂须腕虫，体长60厘米至70厘米，令人们之前所知道的所有须腕动物都变成了小矮子，但是几年之后，人们便又发现了一些真正的庞然大物。

20世纪70年代末有了最知名的发现，当时地理学家们使用深海潜水器“阿尔文”号探索加拉帕戈斯群岛附近一片海域的46水下火山活动，在这里发现了巨大的火山岩烟囱向外喷出富含硫化氢等有毒化学物质的热水。令人吃惊的是，在这种极端的环境中生命是非常丰富的，包括体长达1.5米的巨型管状蠕虫形成的森林。这些动物名为巨型管虫，顶上有血红色的触须，从深海潜水器的窗口望去非常引人注目。巨型管虫、瓣臂须腕虫和其他所有须腕动物有一个很有意思的共同之处：它们都缺少一个非常重要的部位——肠道（内脏）。它们没有口，没有肛门，没有明显的进食路径。这些深海管状蠕虫如何生存的答案藏在它们的身体中。它们体内有一个独特的器官，名为营养体，其中包裹着千百万活的细菌，所有这些细菌都是一个类型的，能够利用通常有毒的硫化氢作为能量来源，制造出食物化学物质。在黑暗的深海中，这些细菌利用“化学合成”制造食物：这种反应与植物的光合作用类似，只是使用的是来自化学键的能量，而不是太阳的能量。深海的管状蠕虫是农民，它们的细菌农场就在自己体内，所以它们并不需要进食。

## 扁形动物门和纽形动物门：一个胖一个慢

并非所有蠕虫都属于环节动物门。扁虫、吸虫和绦虫同属另一个门：扁形动物门。它们与环节动物相差极大，它们体内没有充满液体的体腔。扁虫是非常硬的动物，不会像环节动物那样蠕动或扭动身体，因为它们的肌肉块并没有被分割为独立的节，而且身体没有液体框架，这便意味着没有坚硬的部位供肌肉弯曲。扁虫利用身体边缘小幅度的肌肉收缩来移动，一些体形

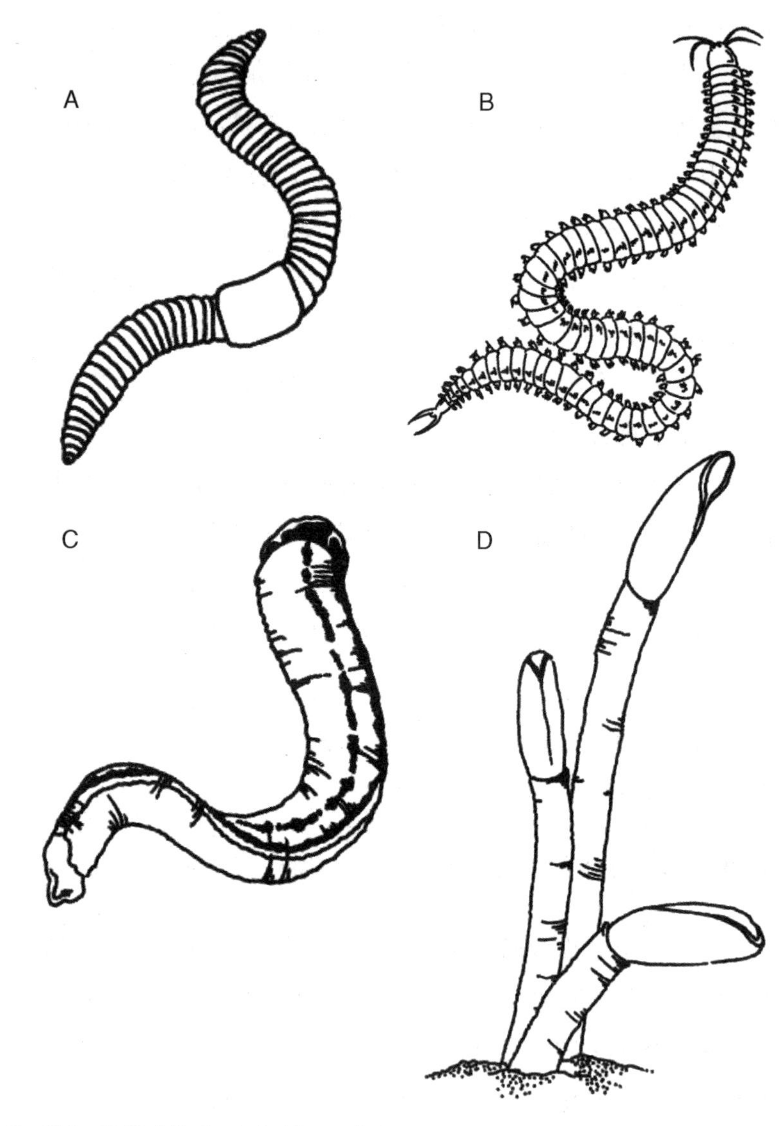


47　图6　环节动物门：A. 蚯蚓；B. 沙蚕；C. 医蛭；D. 须腕动物纲，管虫

极小的物种则是利用表皮细胞探出来的纤毛移动。它们没有循环的血液系统，也没有特化的鳃，这意味着它们只能依靠身体表 48 面的简单透析来为细胞获取氧气，因此这类动物中的大多数皆体形不大，身体扁平。翻开溪涧或河流中的小石头，就能很轻易地发现扁虫的踪迹，它们身体呈椭圆形，体长几毫米到一厘米，移动缓慢而稳定，以藻类和残渣为食。

然而，并非所有的扁形动物都如此无害，好些种类与人类间存在着令人不喜的联系。其中最知名的，可能是裂体吸虫中的代表曼森氏裂体吸虫，如今正感染着超过两亿人。感染症状是多种多样的，但在严重的情况下，裂体吸虫会导致内脏器官的损伤，甚至致人死亡。和很多吸虫一样，寄生裂体吸虫需要两个宿主来完成生命循环，其幼虫在淡水螺体内发育，之后就会进入到河水里，从那里想办法穿透第二个宿主的皮肤，通常是人类。

另一门不分节的蠕虫是纽形动物，俗称丝带蠕虫，通常能在海滩岩石的下面发现。这些蠕虫生活节奏很慢，以非常悠闲的速度缓缓移动，基本上不花费什么力气。它们同样缺少大型的充满液体的体腔，因而能够将身体变形或伸展，形成扭曲的形状，仿佛口香糖扭成串的样子。尽管生活方式懒散，但很多纽虫都是非常贪婪的掠食者，它们的口器很长，上有黏胶或有毒的倒钩，能用来捕食其他的无脊椎动物。大多数纽虫的体长只有几厘米，但一个英国物种巨纵沟纽虫（俗称“靴带虫”），很可能是地球上体长最长的动物。蓝鲸的体长为30米左右，现在已经发现的巨纵沟纽虫的样本自然是超过这一长度的，有说法称其长度是超过50米的。不过即便是这样的怪物，其身体的宽度也从来都只有几毫米而已。 49

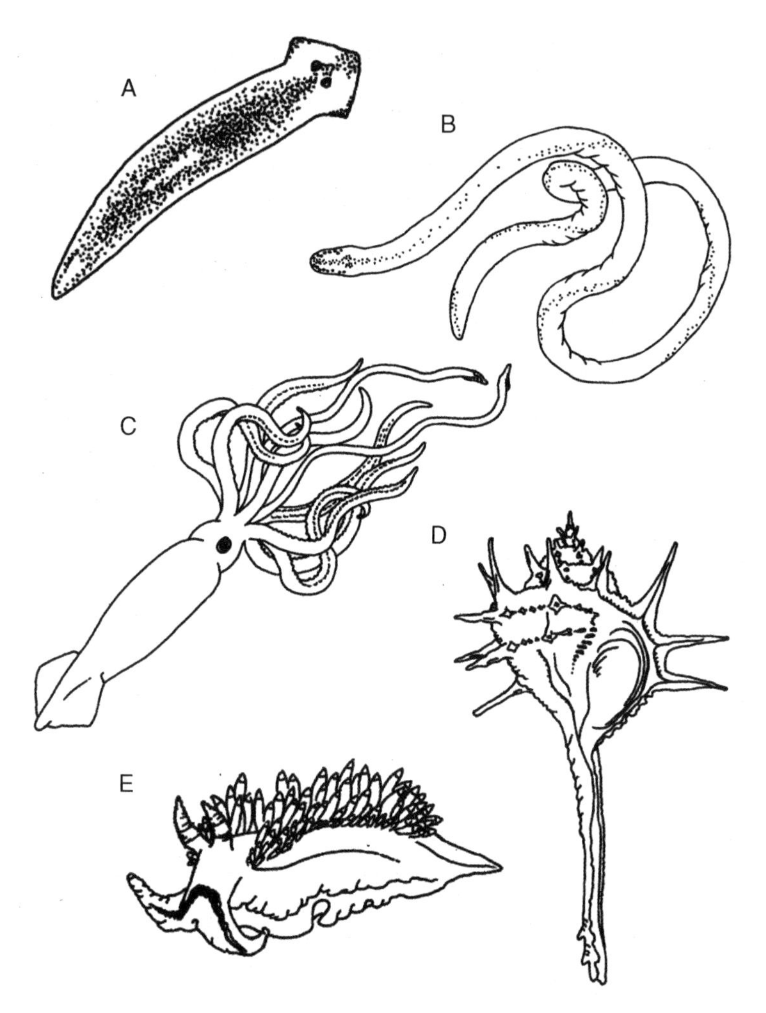


图7　A. 扁虫动物门（扁虫），三角涡虫；B. 纽虫动物门（丝带蠕虫），红纽虫。C—E. 软体动物门：C. 大王乌贼；D. 腹足动物纲，骨螺；E. 翼蓑海蛞蝓

## 软体动物门：从乌贼到蜗牛

一般公认的“最大”的无脊椎动物，属于另一个门：软体动物门。大王乌贼即巨乌贼，是一种体形极大的动物。它的长度

可能怎么都无法和靴带虫相提并论，“仅仅”体长13米，但就绝对体积而言，它要获胜简直易如反掌，确切地说，是易如垂下它的触手。和其他鱿鱼一样，大王乌贼身体短而粗壮，长着八条有吸盘的支腕和两条末端有锯齿状吸盘的触手。这种动物生活在海洋深处，尽管动物学家窪寺恒己近年来成功地拍摄到了活的巨乌贼，不过我们的主要知识仍然来自被冲到陆地上或偶然间被拖网渔船捕获的标本。除了关于可怕的“克拉肯”的古老神话和传说外，至少有一份巨乌贼袭击船只的可信记录，记载在20世纪30年代的挪威海军船只“不伦瑞克”号的日志中。法国三体帆船“杰罗尼莫”号在参加2003年的“儒勒·凡尔纳杯”比赛时也曾遭遇过巨乌贼，一名船员描述说其触手“就和我的手臂一样粗”。而鱿鱼袭击游泳者的故事中出现的，更可能是另一种动物——茎柔鱼，俗称美洲大鱿鱼或洪堡鱿鱼，其身形较为粗短，只有两米。这些大胆的掠食者在捕食时是成群结队的，能快速而凶猛地攻击鱼类和其他游动的猎物。难怪有些潜水者在与成群的茎柔鱼一起游泳时要穿上防护服了。章鱼、乌贼和鱿鱼都是头足纲动物——软体动物的一个主要类别。体形并非它们唯一的出名之处，尤其是章鱼，章鱼可能是所有无脊椎动物中认知发展能力最高的。它们的大脑很大很复杂，视觉非常敏锐，它们可以解决很多难题，比如专门设计用于测试空间记忆的迷宫。

与大多数头足纲动物不同，大多数软体动物都有一个很显眼的壳。外壳由一层特殊的细胞——外套膜——分泌的碳酸钙结块形成，主要功能是抵御捕食者。蜗牛等腹足纲软体动物只在背上有一个单壳，脆弱的内脏都隐藏在壳里。尽管外壳明显很有用处，但腹足纲中的一些种类在演化过程中完全失去

51 了壳。其中很多动物都找到了保护自身的替代方式。为园丁深恶痛绝的陆生蛞蝓会分泌出恼人的黏液，令一些天敌却步，虽然不能吓退全部天敌。而在某些方面，没有壳则是一个很有利的优点，蛞蝓可以在钙含量低的栖息地茁壮成长，蜗牛却不可以。

有一类海蛞蝓，只与陆生蛞蝓和蜗牛有着很远的亲缘关系，却已演化出一种更加令人震撼的防御方式。翼蓑海蛞蝓以海葵等刺胞动物为食，它们不会被海葵蜇伤，反而能够不引发海葵刺丝囊（放射刺丝的细胞器）的攻击就将其缴获。这些亚细胞结构被海蛞蝓回收使用，装载到自己表皮中生长出的组织叶片里。然后，这些水生的腹足纲动物身上便密布着偷来的带有剧毒尖刺的鱼叉，就像海葵和水母一样。软体动物门的第三大类双壳纲动物有两个壳。牡蛎、蚌、贻贝都是此类动物中广为人知的代表，它们的生活方式相似。这种动物隐藏在两个贝壳之间，有着非常精妙的W形的鳃，上面覆盖着成千上万的纤毛，纤毛摆动，便能吸入强劲的富含氧气的新鲜水流。水流还会携带着悬浮在水中的食物微粒，如微小的藻类，卷向它们的口部。



千百年来，软体动物都是人类重要的食物来源。世界各地的沿海地区都曾经发现过巨大的“贝丘”，成堆的被遗弃的古代贝壳，通常长达数百米。除了将软体动物作为食物以外，根据老普林尼和亚里士多德的记载，人类还从软体动物身上提取颜料，特别是染料泰尔紫。这种鲜艳的颜色提取自水生腹足纲动物染料骨螺，将其身体提取物与盐混合后加热能获得这种染料，被用来给古希腊和古罗马贵族的外袍染色。有些种类的软体动物对人类有着有害的影响，如双脐螺，这种淡水螺类是我们之前提过的裂体吸虫的中间宿主。有一种软体动物甚至改变了欧洲历史的轨迹。1588年，西班牙无敌舰队向英国进发，决定去推翻伊丽莎白一世女王的统治。挫败西班牙的确能够彪炳史册，但英国也许并不该把功劳都归于海军上将弗朗西斯·德雷克爵士。启航出征之前，西班牙舰队在里斯本港停泊了好几个月，它的木头船体沾上了一种对木头十分有害的双壳纲动物：船蛆。这种软体动物便是臭名昭著的“船食虫”，它们身体细长，两片壳都退化成位于身体一端的小薄片，用来挖掘它们的食物来源——木头。由于木头上都是洞，战斗还没开始，无敌舰队的实力就遭到了致命削弱。也正是由于船蛆的习性，保留至今的古代帆船数量极少。瑞典的战船“维瑟”号是保存完好的特例，“维瑟”号在1628年首航时沉没于波罗的海，这片海域含盐度低，不适宜船蛆生活。 52

以上所述的门——环节动物门、扁形动物门、纽形动物门和软体动物门均为冠轮动物总门的成员：冠轮动物总门是动物演化树的一个巨大分支，这个名字也很“大”。“冠轮动物”中的“轮”来自担轮幼虫，这是一种特殊的浮游的幼虫，这些门中的一些物种（并非全部）都有这样的幼虫，最明显的是海生环节动物和软体动物。担轮幼虫通常被描述为类似微型的陀螺，不过由于担轮幼虫并不会像陀螺那样转动，因此将其描述为梨形会更形象。冠轮动物的冠是指触手冠，这是一圈不同寻常的进食结构，由支腕组成，形状似冠冕，不过我们前面讨论的几个门中都没有这种结构。事实上，在另外三个不是特别像蠕虫的门中存在触手冠。这三个门分别是生有贝壳的腕足动物门、罕见的帚虫动物门和非常微小的苔藓动物门。苔藓动物聚集生长，如同席子一般，群落通常发现于大型海藻的叶片上。观察运动中的触手冠的最佳方式，是找一片被冲入岩石潮水潭中的海藻叶片，用低倍显微镜或手持放大镜检查其上是否有白色的“海席”。[53] 在被淹没的几分钟内，数百只微小的苔藓动物将开始在海水中摆动它们纤细的触须，寻找食物颗粒。我们通过DNA序列的比较才发现，生有触手冠的动物和有担轮幼虫的动物属于动物界的同一分支。这一大类即冠轮动物总门，它与我们接下来要介绍的另一大类两侧对称动物——[54] 蜕皮动物总门是姐妹。

第七章

# 蜕皮动物总门：昆虫和线虫

大致来说，所有物种都是昆虫。

——罗伯特·梅，

《自然》，324（1986）：514-515

## 昆虫：陆地的主人

没有人知道有多少种昆虫。对种类数的估计结果相差很大，从几百万到三千多万不等。至少有80万个物种已经被记录并正式命名，但即便这个数字也不准确，因为始终都没有汇编过总清单。那些已经被记录的物种的地理分布、生态及行为习性，很多都还是未知的。但是为什么会有这么多种昆虫呢？这并非一个容易回答的问题，不过很可能有以下两方面的原因：能轻易适应多种生态位和食用植物的身体构造，以及植物物种的多样性，特别是热带地区的植物。另外，在大约四亿年前，在动物演化的初期，陆生的昆虫就已经离开海洋，因而有了足够的时间随着陆生植物的演化而发展出巨大的多样性。

昆虫是最重要的陆生动物。它们是节肢动物门的一部分，

55 和所有节肢动物一样，昆虫有坚硬的外骨骼和用于运动及进食的有关节的肢体，为了生长，昆虫会不断地把外骨骼蜕掉。尽管它们的祖先生活在大海里，但昆虫已经演化出一系列的适应性改变，以便能够生活在极易出现极端温度和严重缺水的恶劣环境中，这个恶劣环境也就是我们所说的陆地。不论是在大海里还是在陆地上，外骨骼都为身体提供固有的支持，而对昆虫来说，表皮的最外层上还有蜡，已经可以防水，也可以有效地阻止水分蒸发出身体引起的干燥。这解决了一部分问题，但依然有两个过程容易流失水分。第一，动物需要获取氧气，排出二氧化碳，气体扩散的物理学原理表明，通过湿润的表面效率最高。为防止将湿润的表面暴露于外部环境，从而破坏防水的外骨骼的作用，昆虫演化出了一系列精巧的表皮之下的"气管"，这些管道弯弯曲曲，分支众多，从身体表皮上可以关闭的气孔直通到内部组织。气孔顶部没有角质层，昆虫可以按自己的需要在合适的地方进行气体交换。第二，所有动物都需要处理含氮的废物，这些废物是蛋白质代谢过程中产生的，但是对细胞有毒性。很多动物，包括人类，通过稀释废物然后排出尿液来解决这个问题，但这样很浪费水分。而昆虫使用了不同的代谢途径，产生能结晶为固体的尿酸——而非可溶性氨或尿素——然后利用高效的腺体，在排泄之前先将水分吸收。由于尿酸是无毒的，很多昆虫会将一些尿酸储存在特化细胞里，还有一些动物将其利用起来。欧洲粉蝶等粉蝶属的蝴蝶的白粉就是由储存在翅膀鳞片中的尿酸生成的。

节肢动物门的所有成员都有一个明显特征：分节。身体的

大多数重要组成部分，包括与运动有关的肌肉和神经，按照一系列模式连续重复，延伸至整个身体，就仿佛身体被分成了一系列的单元。环节动物门分节的蠕虫的结构方式与此类似，但是——与长久存在的观点相反——这两个门完全不是近亲。环节动物门被分在冠轮动物总门，节肢动物门则在蜕皮动物总门。节肢动物的分节同样影响了坚硬的外骨骼的生成，因此在节与节之间需要表皮较为柔软的关节部位，以便身体可以扭动和移动。如果没有关节，动物将被包裹在一层不可移动的盔甲中。昆虫的分节模式已经以一致的方式发生了变化，而这种变化可能也激发了昆虫的显著适应性。昆虫并不全身都是一系列几乎相同的节，而是一组一组的节组合在一起，合成三个基本的节段，被称作"体段"。首先是头部，由六七个节合在一起构成，包含主要的神经中枢、感觉器官和有关节的进食结构。柔韧的脖子之后是胸部，由三个节构成，每一节有一对有关节的腿。大多数昆虫胸部的第二节和第三节（T2和T3）各有一对翅膀。最后是腹部，由八到十一个节构成，虽然不太坚硬而且没有腿，但里面包裹着动物大部分的消化器官、生殖器官和排泄器官。就功能而言，头部负责进食和感觉，胸部负责运动，腹部负责新陈代谢和繁殖。这些功能的分开使身体的每一个部分都能得到优化。

56



## 统治天空：翅膀与飞行

在地球的生命历史中，动力飞行只演化了四次——分别表现在鸟、蝙蝠、翼龙和昆虫身上。唯一能飞行的无脊椎动物是昆虫。它们也是地球上数量最丰富、种类最多样的飞行者。飞行是理解昆虫成功的绝对关键。有趣的是，尽管有点令人费解，但昆虫演化出了两对翅膀。两对翅膀比一对翅膀好吗？毕竟，鸟和蝙蝠都只有一对翅膀，不过，某些可能与鸟类的祖先有亲缘关系的恐龙的“上肢”和“腿”上都有羽毛。翅膀数量之所以不同，可能与脊椎动物胚胎发育模式的限制有关。有些证据证明，脊椎动物如蝙蝠和鸟，只能有两对肢体，如果一对用于行走，那么就只有一对能用来飞行。昆虫的翅膀不是从腿演化来的，因而没有这样的限制存在。因此，胸部的第一节（T1）只有腿，胸部的第二节和第三节（T2和T3）既有翅膀又有腿。两对翅膀使得昆虫能够采用的飞行方式更加多样。

昆虫分成差不多30个目，包括蚱蜢（直翅目）、蜻蜓（蜻蜓目）、蜉蝣（蜉蝣目）、竹节虫（竹节虫目）、蠼螋（革翅目）、蟑螂（网翅目）、陆生和水生的臭虫（半翅目）、跳蚤（蚤目）等。其中四个最大的目，包含了超过80%的已经被记录的物种，这四个最大的目包括甲壳虫（鞘翅目），蝴蝶和蛾子（鳞翅目），蜜蜂、黄蜂和蚂蚁（膜翅目），以及苍蝇（双翅目）。每一个目都有惊人的多样性，每一个目都以不同的方式改变了自己的翅膀。



鳞翅目昆虫有两对发育完全的翅膀。有些蛾子有刺或突起，将前翅和后翅连在一起，但鳞翅目中的多数昆虫在飞行过程中，均可以独立移动并分别控制翅膀。翅膀的形状也千差万别，羽蛾的翅膀上有羽毛状的探出物，生活在南美洲的袖蝶生长着细长的桨叶状翅膀，而燕尾蝶则有宽阔的滑翔翅。

蛾子和蝴蝶看起来可能外形脆弱而短命，但有些物种是非常强健而又长寿的。黑脉金斑蝶俗称帝王蝶，在墨西哥中部的群居栖息地越冬，然后进行飞跃北美洲的集体迁徙。每只蝴蝶将飞行数百公里，只需几代蝴蝶，后代便能够到达北方的加拿

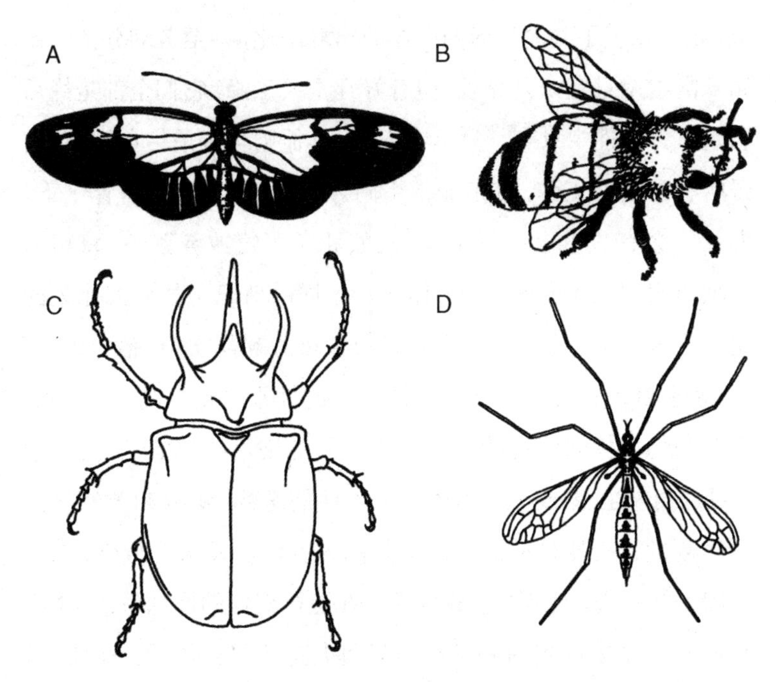


图8 昆虫，四大类：A. 鳞翅目，蝴蝶、袖蝶；B. 膜翅目，蜜蜂；C. 鞘翅目，甲虫、南洋大兜虫；D. 双翅目，大蚊

大——距离冬季的栖息地4 000公里远。小红蛱蝶又名浓妆夫人蝶，同样以迁徙行为而闻名。欧洲的博物学家几乎都不会忘记1996年和2009年大群的小红蛱蝶从非洲的阿特拉斯山飞来，一路向北飞掠过欧洲，并沿途繁殖，最终到达苏格兰和芬兰这样遥远的北方。

蚂蚁、蜜蜂和黄蜂——膜翅目——也有两对翅膀，但这两对翅膀通常被后翅上的一排钩子紧紧地固定在一起。大多数种类都已经能适应快速而有控制的飞行，因而蜜蜂可以盘旋飞行，或急速冲入小小的空间以收集花蜜，马蜂能瞬间捕获猎物，寄生蜂能落在毛毛虫附近，然后将卵产在毛毛虫体内。主要在膜翅目中，我们发现了共同生活的个体的集群演化，甚至劳动分工。比如一窝蜜蜂只有一只蜂后，但有好几千只工蜂，它们全都是蜂后的姐妹。雌性中只有一只负责产卵，其他则负责采集食物、清洁居所和防御等工作，这是非常不同寻常的，需要详细解释一下。为什么会有成百上千只工蜂、工蚁和工黄蜂放弃繁殖，将精力奉献出来，帮助另外一个个体呢？这样的情况是如何演化来的呢？答案并不简单。一个流行了很多年的解释以“单倍二倍体”（在膜翅目动物身上发现的不同寻常的性别决定机制）为基础。许多动物雄性和雌性的分别源于一条性染色体，如人类的性别取决于X染色体或Y染色体。但雄性的蜜蜂、蚂蚁和黄蜂的染色体数量只有雌性的一半。这是因为受精卵发育为雌性，未受精卵不会死去，而是发育成雄性。在这样奇怪的性别系统中，姐妹们——如工蜂和蜂后——在基因上是非常相似的。事实上，雌性的蚂蚁、蜜蜂和黄蜂，与其姐妹的相似性，要大于与其后代的相似性。这也许能说明，从演化层面上说，雌性之间的协作是更有利的，因为通过帮助蜂后和蚁后，工蜂和工蚁也顺带促进了自己基因谱系的存续。然而，这个经常被人引用的解释所存在的问题是，根据单倍二倍体理论，雌性只与它们的兄弟们之间存在微弱的亲缘关系，这就抵消了遗传优势。蚂蚁、蜜蜂和黄蜂的这种社群性的演化起源，可能与独特的性别机制关系不大，反而更可能是为了与亲缘动物共享防御资源，以及为了繁殖体系的需要而长期持续地照顾幼虫。

膜翅目的前后翅连接在一起，意味着两对翅膀具有相同的机械特性，就像一对翅膀。昆虫中最大的两个目演化得要更进一步，只用一对翅膀来飞行。鞘翅目即甲虫类，只用后翅飞行。

双翅目即蝇类，只用前翅飞行。这两大类当然都是从使用两对翅膀飞行的昆虫演化来的。甲虫祖先的前翅已经演化成坚硬的翅鞘，在后翅不使用的时候覆盖在后翅上面起到保护作用。这种改变为甲虫开启了新型的生态环境：它们可以在泥土里挖掘，钻到植物的种子中，或在腐朽的木头中打洞，而不会伤害到轻薄精巧的用于飞行的翅膀。一直以来，甲虫巨大的多样性吸引了一代又一代的生物学家，查尔斯·达尔文年轻时就曾是鞘翅目的迷恋者。达尔文极为喜欢一种体形很大的圣甲虫，在《人类的由来》一书中，他这样写道：

> 如果我们想象一只有光滑的青铜色外壳和巨大而复杂的触角的雄性南洋大兜虫被放大到一匹马的大小，哪怕是一只狗的大小，它也将是这个世界上最壮观的动物之一。

双翅目苍蝇的后翅已经演化成小小的棒状的“平衡棒”。平衡棒在飞行过程中上下振动，与前翅的拍动不同步，它们是复杂的感觉反馈系统的一部分。如果苍蝇的身体向一侧倾斜，平衡棒则可能在陀螺仪的惯性作用下继续维持它们原本的震动面，平衡棒根部的感觉器官能察觉出平衡棒与身体位置之间的角度变化。因此，苍蝇能持续接收到关于其在空间中的精确方位的连续信息。所以，苍蝇是所有昆虫中最敏捷的，能够以最惊人的速度和准确度盘旋、急速前进或转头。

双翅目昆虫数量众多，有数万种之多，其中有很多种都对人类生活有巨大的影响，其中包括蚊子，它们会传播疟疾寄生虫或携带导致黄热病和登革热的病毒。每年有超过100万人死于由

61 蚊子携带传播的疾病。一些蝇类对人类有益，比如可以作为授粉者。黑腹果蝇这个物种在医学研究中发挥着重要作用。100多年以来，这种果蝇一直是遗传学研究中最受欢迎的“模型动物”，它体积小，易于大量繁殖。使用果蝇所做的研究增进了我们对基因功能和交互作用的了解，而这些知识与很多人类疾病相关，包括癌症。

## 其他的节肢动物：蜘蛛、蜈蚣和甲壳纲动物

除昆虫之外，还有三类现存动物属于节肢动物门。其中两类为螯肢亚门和多足亚门，它们已经成功侵入陆地。第三类甲壳纲动物主要是水生物种，但也有一些陆生物种。螯肢动物包括蜘蛛和蝎子等，尽管大部分都生活在陆地上，但最初都是起源于海洋的。它们的结构和对陆地生活的适应与昆虫极为不同，很显然，螯肢动物和昆虫侵入陆地的过程是各自独立的。

多足亚门中，蜈蚣和马陆是最知名的种类。这些动物有一个特化的头，后面是一系列分节，由有关节的腿支撑。蜈蚣的身体由一圈圈柔韧的表皮分隔，这使得它们可以十分迅速地扭动身体、转身和奔跑。蜈蚣是掠食者，会积极主动地追逐和捕捉猎物，用凶猛的“毒爪”（从一对前腿演化而来的巨大毒牙）攻击。而马陆则不同，它们主要以木头和腐叶为食物，行动反应要慢很多。它们也没有毒爪，很多种马陆的节是相互紧扣的，这些马陆可以像慢动作的攻城锤一样缓缓地穿过泥土或腐烂的植物。蜈蚣有100条腿，马陆有1 000条腿，这种说法并不是真的，不过，如果真要说到腿的数量，有些特别之处还没有被完全理解。蜈蚣最少有30条腿（2×15），最多会有382条腿（2×191），但奇怪的

是，蜈蚣用于行走的腿的数量总是奇数对，也就是说，在不算毒爪的情况下，并没有一种蜈蚣刚好有100条腿。即便是分节数量不同的物种，个体间腿的差别也总是两对的倍数。而马陆则有另一个奇怪之处。从上方俯视时，马陆的每一节看似有两对腿，

62



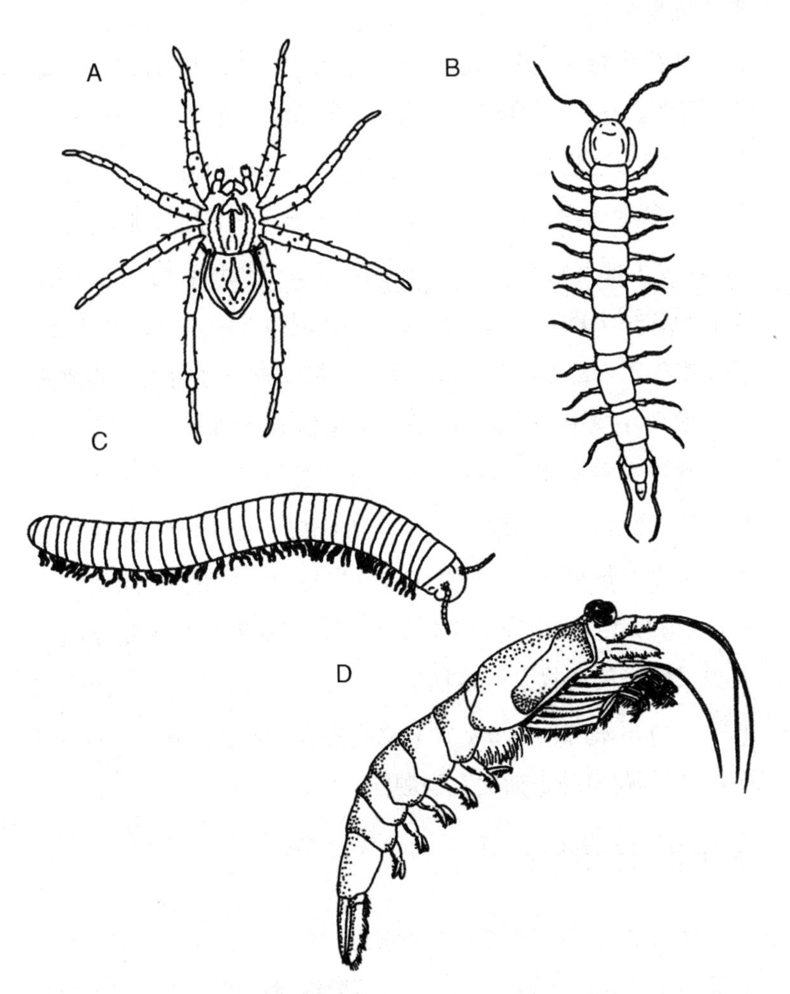


图9 节肢动物门：A. 螯肢亚门，蜘蛛；B. 多足亚门，蜈蚣；C. 多足亚门，马陆；D. 甲壳纲，磷虾

因此也许可以得出一个观点：在演化过程中，它们的节两两地连在一起，形成了“双节”。

然而，这种模式从下面是看不到的，近年的基因表达研究显示，顶部和底部的节之间的分界是不同的。马陆的节不是简单的重复单位。

和昆虫相似，多足动物同样利用气管将氧气输送到身体组织，同样有不分叉的肢体。蜈蚣、马陆和昆虫的头部结构也非常相似。因为这些相似性，在长达100多年的时间中，生物学家们都认为多足动物和昆虫是非常近的近亲——同属节肢动物门。然而，分子证据则指向了不同的结果，并强烈地表明：昆虫实际上和甲壳纲更近。昆虫可能属于甲壳纲的下层。甲壳纲动物主要是水生的，这就意味着昆虫和多足纲动物是分别侵入陆地的，为适应新环境中的生活，都独立演化出器官和不分叉的腿。甲壳纲动物种类众多，包括很多人类熟悉的动物，如螃蟹、龙虾、虾，甚至还包括一些寄生的物种，如鱼虱。很多甲壳纲动物在生态意义上非常重要，如在海洋浮游生物中大量存在的桡足类动物，或须鲸赖以为食物的大群的磷虾。其中最不同寻常的，也是人类最熟悉的动物之一是藤壶，它们最初的生命形式是可以在海洋中自由游动的幼虫，后来就固定在岩石上，头朝下挂在上面，余生都蹬着腿去捕捉食物颗粒。

## 水熊和天鹅绒虫

有两个与节肢动物亲缘非常近的门，几乎名列所有动物学家最喜欢的动物之列。它们是用显微镜才能看到的缓步动物门和生活在森林中的有爪动物门。这两类动物都有短而粗的腿和

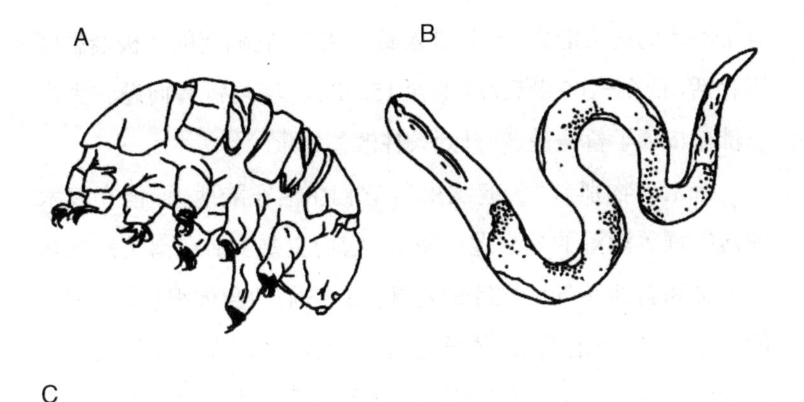


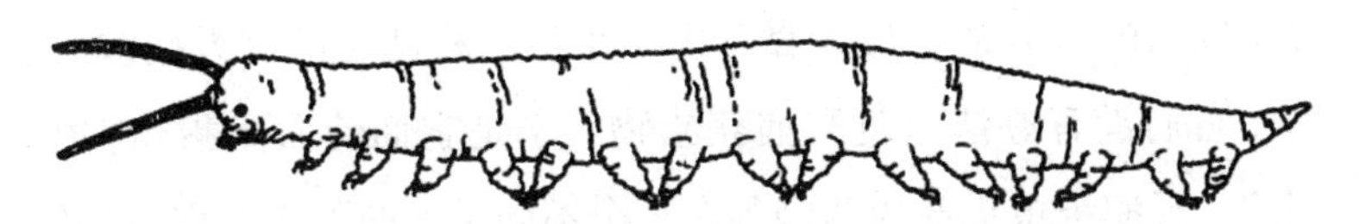


图10 A. 缓步动物门，水熊；B. 线虫动物门，蛔虫；C. 有爪动物门，天鹅绒虫



柔软的外皮，而不像昆虫及蜘蛛等节肢动物那样有坚硬的、有关节的腿和坚硬的外骨骼。缓步动物别名“水熊”，长度不足一毫米，在水面的潮湿苔藓或地衣上可以发现它们的踪迹。

64

通过显微镜观察可以看到，它们粗胖的身体和懒散的步态让它们看起来真的很像缩小版的熊，不过是有八条腿的熊。除了外形讨所有人喜欢，缓步动物还因一项了不起的能力而知名，它们能忍受极端的环境条件。如果它们的栖息地慢慢地变得干燥，缓步动物就能分泌出一种蜡质外皮，把腿收起来，整个身体变得就像一个小圆桶。然后，它们会减少对氧气和水的消耗，渐渐进入一种接近假死的状态。这种状态被称作隐生状态，缓步动物能以这种状态活上好多年。有人说它们能活100年，不过，以近年的研究来看，这么长的存活时间不太可能。缓步动物还有了不起的适应能力，一旦进入隐生状态，它们便能在极端温度下存活，低至-200摄氏度，高至150摄氏度，均没有问题。生命仿佛被按下了暂停键，一直到条件改善为止。

有爪动物别名“天鹅绒虫”，它们生活在陆地上，能在腐木或落叶堆等潮湿的栖息地中找到，它们主要生活在南美洲的热带丛林和新西兰气候微凉的丛林中。它们的身体很软，有细长的绒毛，有些像毛毛虫，体长几厘米，长着20对左右又短又粗又软的腿。尽管天生移动缓慢，但大多数天鹅绒虫实际上是捕猎者，以白蚁和其他昆虫为食。由于小昆虫移动速度比天鹅绒虫快，它们没有办法追赶和捕获猎物。于是它们转而向猎物“射击”。天鹅绒虫头部的两侧各有一个不同寻常的附属器官，被认为是从腿演化而来的，用来向预定目标发射一股股黏液。这种胶能缠住昆虫猎物，然后天鹅绒虫便能悠闲地享用大餐。为了不浪费有价值的能量，以蛋白质为主要成分的胶会被一起吃掉。

## 蜕皮的蠕虫

线虫又名蛔虫，是节肢动物门最远的亲戚。它们不分节，没有外骨骼，没有腿脚。而且正如它们的名字，它们只是蠕虫——细长而有弹性。然而，1997年以来，越来越多的DNA序列证据逐步积累，表明了线虫在动物界中的确处于与节肢动物、水熊、天鹅绒虫非常接近的位置，此外还有一些不太知名的动物，如只能用显微镜才能看到但名字非常奇妙的动吻动物，别名为“泥龙”的动吻虫。对这一发现，大多数动物学家都非常吃惊，因为过去的解剖学研究从来都没有这方面的暗示。但实际上，所有这些动物都有一个相同的基础特征——它们在成长过程中都会蜕皮。节肢动物有坚硬的外骨骼，需要不断地蜕掉，下面的身体才能长大，这个过程被称作“蜕皮”。水熊和天鹅绒虫（以及毛毛虫等昆虫的幼虫）的外皮相对柔软而有弹性，但它们也蜕皮，因为外皮的分子结构并不适合扩张。线虫的外皮构造非常复杂，由紧紧包裹的蛋白质纤维构成，围绕身体形成一层层密集紧实的弹簧。这些也必须被蜕掉，线虫才能生长。DNA证据显示，这些门是有亲缘关系的，这类动物必须有自己的名字。安娜·玛丽亚·阿吉纳尔多、詹姆斯·莱克及其同事是最早观察到这种关系的生物学家，他们将这类动物命名为“蜕皮动物总门”。 66

线虫的内部结构也非同寻常。和其他很多蠕虫一样，它们的身体内有充满液体的空间，但它们把这些液体维持在很高的液压水平，大约是其他虫子体内液压水平的十倍。这种内部的压力将线虫的组织和外皮向外推动，使身体呈现出环形的横断面。因此在英文中，它们被俗称为“圆虫”（roundworm）。线虫的另一个特别之处是它们身体上的肌肉按照从头到尾的方向（纵向）排列，它们没有环绕身体的环肌。大多数其他蠕虫，如蚯蚓、沙蚕、纽虫，都有两种类型的肌肉，能够相互对抗地收缩，从而改变身体的形状，让蠕虫得以爬行或挖掘。线虫的肌肉只能纵向地收缩，那么它们是怎样扭动身体或移动的呢？答案在于充满高压液体的体腔和具有弹性的外皮，它们与肌肉抵抗，令虫子的身体可以快速波动。它们的运动不像蚯蚓或沙蚕那么协调，部分是因为它们的肌肉排列不同寻常，部分是因为线虫是不分节的，不能轻易地让身体的不同部分向相反的方向运动。线虫以一种拍打的动作进行移动，这种移动方式在游泳时效率不高，在另外一个地方却很完美，那便是它们喜欢的住所——其他东西的内部。很多种类的线虫都生活在泥土或腐烂的蔬菜中，腐烂的水果会是它们的聚集地。甚至有一种以酵母为食的“啤酒杯垫线虫”，其学名为全齿复活线虫。很多其他种类寄生在植67物和其他动物体内。人类也无法对它们免疫，很多严重的疾病都是由寄生线虫引起的，包括河盲症、几内亚龙线虫病、弓蛔虫病和象皮病。

1914年，有“线虫学之父”之称的内森·奥古斯塔斯·科布曾经以诗意而又有些夸张的笔法描述了线虫在其他生物体内的生活习性：

> 如果宇宙中除了线虫以外的所有东西都被清除掉，我们的世界依然是隐约可以辨识的；如果可以将这个世界作为没有了肉体的灵魂来研究，我们会发现山脉、丘陵、谷地、河流、湖泊和海洋都由一层薄薄的线虫来勾勒。城镇的位置是可以辨别出来的，因为只要人群聚集，就会相应地聚集着一群线虫。

一类格外细长的虫子和线虫的亲缘关系非常近，和线虫有很多的相似性，被称作线形动物。尽管身体的宽度很少超过一毫米，但其体长通常能长达50厘米至100厘米。和线虫动物类似，线形动物也有坚韧的表皮，会随着成长不断蜕皮，而且它们也仅有纵向的肌肉。而和线虫不同的是，它们什么都不吃。至少成虫什么都不吃，其肠道已经萎缩退化为一团残存物。未长成的线形动物当然会进食，在其宿主的身体组织内进食，它们的宿主是节肢动物，可能是蚱蜢、蟑螂或淡水虾。这些幼虫在那里成长和蜕皮，身体变长，直到体形大到不能再待在宿主动物体内，它们就会冲破宿主的身体爬出来，只余下这不幸宿主的残骸。成虫必须生活在水中，如果宿主是蟑螂这样的陆生动物，寄生虫就会设法控制宿主的行为，迫使它们向水移动，扎入水中，等待可怕的死亡。寄生在淡水虾等水生宿主体内的线形动物通常被称为“马毛虫”。在这些动物真实的生命循环为人所知之前很久，乡下的人偶尔注意到在马喝水的干净水槽中游着一种细细长长的虫子，这些虫子是凭空出现的，前一天还不存在。于是便有了马尾巴上的毛落到水里有了生命的传说。真相尽管没有这么神秘，但要更加恐怖，这些巨大的蠕虫是寄生虫，本来寄生在水中不显眼的小虾身上，然后破体而出。

68

69

第八章

# 后口动物总门（上）：海星、海鞘和文昌鱼

我在这里还要向棘皮动物致敬，这是一个高贵的群体，专为迷惑动物学家而设计。

——莉比·海曼，

《无脊椎动物（第四卷）》（1955）

## 来自胚胎的线索

棘皮动物一直被公认为地球上最奇怪的动物，这有充足的理由。海星、海胆、蛇尾、海参和海百合——属于棘皮动物门的五类动物——之间有很多共同之处，但和其他任何动物都没有多少相似之处。它们的构造方式似乎与地球上的其他一切都不相同，即便如此，很早之前，人们就认识到它们与我们人类同属动物界的同一个分区：后口动物总门。揭示这种关系的最初证据是在胚胎中发现的。

大多数动物生命最初的形式都是一个单细胞，即受精卵。受精卵然后分裂成两个细胞，再分裂成四个、八个、十六个细胞，

如此继续下去。尽管这看起来非常简单，但如果对比不同的两侧对称动物，能发现数个不同的模式。最常见的两种是螺旋卵裂和辐射卵裂。透过显微镜观察发育中的胚胎，这两种模式的区别一目了然。螺旋卵裂模式中，四个细胞分裂成八个细胞时，新生成的四个细胞最后会位于四个老细胞中间缝隙的上方。如果你尝试把四个橘子摞在另外四个橘子的上面，这正是你会采取的方式。辐射卵裂的新细胞直接位于四个老细胞的上方，如果用橘子来尝试这种方式，会需要相当高超的平衡技巧。 70

无论是螺旋卵裂还是辐射卵裂，之后的每次细胞分裂都会重复相同的模式，直到最终形成一个由细胞构成的中空的球体。然后以球体表面的某一个点或某一条缝为起点，有些细胞会向内移动，仿佛一根手指或一只手正被推入一个充了气的气球一般。细胞层向内折叠形成的凹口被称作胚孔，而最初的细胞球则被称作囊胚。随着胚胎进一步发育，这个向内缩进形成的管道最后会变成肠道。在这个过程中，螺旋卵裂和辐射卵裂这两种模式会出现第二个显著不同。螺旋卵裂的动物的胚孔将肠道末端作为口部，但更常见的是，胚孔会是裂隙状的，中间部分闭合，两端各有一个开口，分别是口部和肛门。辐射卵裂的动物的胚孔会在胚胎的后端，也就是形成肛门的位置。随着最初的肠道向深处延伸，口部会在发育的胚胎的另一端完全独立地显露出来。因此，有很长一段时间，螺旋卵裂的动物被称作“原口动物”，“原口”的意思便是“最初的口”，因为人们观察到口部由胚胎发育形成的最初的开口发育而来。而辐射卵裂的、胚孔在后端的动物，则被称作“后口动物”，“后口”指的是“第二个口”。不可避免的是，并不是所有的动物都属于这两种明确的模式之一，特别是那些胚胎中有大量影响细胞分裂的卵黄的动物。

这种区别是由卡尔·格罗本于1908年提出的，但时隔一个世纪，再利用它就需要注意了。冠轮动物总门和蜕皮动物总门这两个两侧对称动物的总门是根据分子分析确定的，其中包含了所有原口发育模式的动物，但还包含了很多其他动物。比如71 蜕皮动物总门中的昆虫和线虫并不具有螺旋卵裂模式，也不具有辐射卵裂模式。尽管如此，我们今天依然经常用“原口动物”来总称冠轮动物和蜕皮动物，大家知道这只是一个方便的名字，而非一个统一的规则。同样令人困惑的是，现在被称为“后口动物总门”的演化群体，也是根据分子分析界定的，与格罗本最初的指涉也略有不同。现在被界定为“后口动物”的群体仅仅包含了一些具有辐射卵裂模式和第二个口部形成过程的动物，并非全部。也许最好的办法是废弃原来的旧称，但命名也不总是讲究逻辑的。只要记住一点，并非所有的原口动物都是原口发育的，有些被分类到原口动物的动物实际上是后口发育的。相应地，不是所有具有后口发育过程的动物都是“真正”的后口动物。按照当今的定义，在后口动物总门中，只有三个主要的门，可能还可以加上一两个“小”门。这三大类是棘皮动物门、半索动物门和脊索动物门。



## 与5有关的生命

将一个苹果拦腰切开，你会看到一个包含种子的五角星。仔细观察一朵野蔷薇，注意它的五个花瓣。无论是水果、花瓣还是叶子的模式，5这个数字在植物界中无处不在，是很多变异和适应性改变的基础。而动物则刚好相反，对数字5充满排斥。可能有人会说我们有五根手指，但因为我们有两只手，所以真实数字该是10（如果算上所有手指脚趾，该说是20）。对于有中心对称平面的动物来说，数字5是不怎么适用的，这一点我们在动物界大多数地方都可以看到。2、4、6、8随处可见，但5不是。刺胞动物门中水母那样的基础动物，并没有明显的左右对称的平面，但即便这些动物，通常也具备四重对称，而非五重。 72

棘皮动物门与众不同。这一整个门的演化都被数字5主宰着。这种模式最容易在海星和蛇尾身上看到，它们是海岸和潮下带常见的两种无脊椎动物，有五条腕从中心区域或圆盘呈辐射状探出。海星的腕是非常坚硬的。它移动时的样子像是在海床上滑行，这是通过使用从底部伸出的数千只微小"管足"实现的。管足的运动由体内一系列分布广泛的充满液体的管道驱动，这些管道是棘皮动物特有的，被称为水管系统。蛇尾乍看起来与海星相似，实际上却不相同，它们的五条腕更纤细更有弹性，能够通过抓取和拉动来协助动物的移动。这两类动物在生态学上也非常不同，特别是从扇贝的视角来看。蛇尾以残渣和岩屑为食，通过位于中央圆盘中间的一个向下开口的口部摄取微小的碎粒。而大多数的海星却是贪婪的掠食者。它们移动缓慢，但如果猎物根本不动，再慢都不是问题。大多数海星会猎食贻贝、生蚝、蚌等双壳类软体动物，这些动物生活在两片严丝合缝的壳中，不爱活动。尽管双壳类软体动物通常能免于掠食者的攻击，海星却是它们最可怕的噩梦。当捕猎的海星遇到蚌这样的猎物时，便会将自己的腕缠绕在蚌体上，用吸管一样的管足紧紧吸住表面，然后用力拉。只要两个壳中间出现一道细小的裂缝，海星就会（相当冒险地）将自己的一部分胃翻出来，从自己的口里探出，挤入裂缝当中。胃能分泌分解蛋白质的酶，弱化蚌的肌肉，然后壳会被撬开得更多。最后，蚌的身体就整个暴露出来，被海星吞食掉。因此，双壳类软体动物中少数会游泳的物种，如皇后海扇蛤，只要闻到一丝海星的气味，就会立刻逃之夭夭，这也就不足为奇了。

73 浑身密布防御性刺的海胆和柔软细长的海参也都属于棘

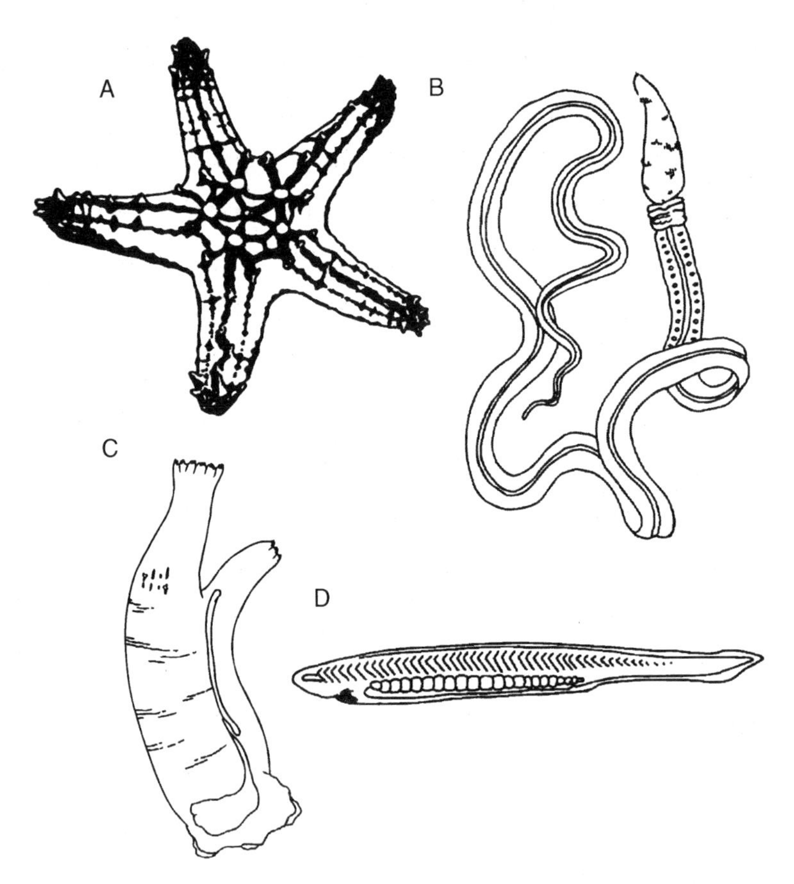


图11 A. 棘皮动物门，海星；B. 半索动物门，柱头虫。C—D. 脊索动物门：C. 成体海鞘，玻璃海鞘；D. 文昌鱼

皮动物。如果只是无心一瞥，它们身上5的痕迹不是那么显眼，但当然还是存在的。这两种动物身体周围都有五个区域长着管足，这表明这些动物是从类似海星的祖先演化而来的，只是它们的腕被折叠到了身体的其他部位上。这个门中的第五类动物是海百合，它们是滤食性动物，身体上方有一个口，口的周围是五根柔软的腕组成的冠。它们的冠有时位于一根长茎上方，特别是深海物种。由于管足和向上开口的口部，海百合与海星和蛇尾的上下方向是相反的。



棘皮动物身上这种五角状对称的演化起源是非常有趣的。五角状对称显然是从左右两侧对称演化而来的，有三个关键原因可以证明这一点。第一，棘皮动物的幼虫和很多其他海洋动物的幼虫一样，是两侧对称的。直到它们摆脱浮游状态，经历变形，才会出现五角模式。第二，棘皮动物的化石中具备所有形式的对称，包括两侧对称，这表明五角状的模式是演化过程中相当晚的时期才偶然出现的。第三，也是最关键的一点，在动物演化树上，棘皮动物门能很好地从属于两侧对称动物，这表明，如今活着的所有两侧对称动物均是从共同的祖先演化而来的。



## 半索动物门：散发恶臭的蠕虫

几年前，我想采集一只柱头虫。柱头虫是一种非常古怪的不分节的蠕虫，属于半索动物门，这个门在演化树上的位置与棘皮动物门很近。它们胚胎的早期发育过程与棘皮动物十分相似，同样具有螺旋卵裂和后口模式，它们的幼虫偶尔会在浮游生物的样本中发现，很容易与棘皮动物的幼虫混淆。在那之前，我从没有见过野生的柱头虫，为了某个研究项目，我需要一个样本。但写信给一些海洋生物学家求助时，收到的一个答复令我迷惑不解。那位生物学家告诉我，他也从来没有在英国见过野生柱头虫，但他十分肯定它们存在于某片特定的海滩上，因为他肯定他闻到了。当然，我绝对不可能相信单纯的气味信息的证据，直到我自己采集了一次柱头虫。大多数的半索动物都只有几厘米长，隐藏在沙地或泥地的洞穴中，从表面的海水中过滤食物微粒。它们依靠咽喉部位的一系列裂缝（咽裂）来采集食物微粒，水从裂缝经过，类似从鱼的鳃部经过。实际上，这个组织很可能和鳃是同源的，也就是说半索动物和脊椎动物的共同祖先在咽喉部位有裂缝，用于获取食物和氧气。很多柱头虫确实有一股强烈的药味，非常像碘酒的味道，这味道的来源是一种有毒的化学物质——2,6-二溴苯酚，这种化学物质在它们的皮肤中浓度极高。这种化学物质的功能并不完全清楚，可能是用于防止掠食动物吞食它们，或者是抑制洞穴内细菌的滋生，也可能两个作用兼而有之。无论其适应性功能到底是什么，这个气味在衣物、手指上经久不散，一旦闻过，永远都不会忘记。

75

柱头虫并非半索动物门中唯一的成员。一同演化的是一类名为羽鳃纲的动物，这是一种住在管子中、顶部有一圈触手的微小动物。你如果不清楚该到什么确切的位置去寻找，是很难遇到它们的。英国最知名的羽鳃纲动物是杆壁虫，体长不足一毫米，在一种软体动物欧洲蚶蜊废弃的壳的内表面能很明显地看到它们细小的白色管子——尽管如此，也只能在英国海岸的某些特定地区才能找到它们。在百慕大和斯堪的纳维亚半岛的峡湾中也发现了其他种类，但这些种类的生物属性依然不为我们所知。1876年英国皇家海军舰艇“挑战者”号在麦哲伦海峡的海床上发现了羽鳃纲一个很不同寻常的属：头盘虫属。头盘虫有咽裂，因而证明了羽鳃纲与柱头虫是近亲，同属半索动物门。半索动物门的第三个属是无管虫属，这个属十分特别，因为它似乎并不生活在管子中。关于无管虫的生物属性，我们所知甚少，因为到目前为止，我们只有43个无管虫样本，而这些全都是由日本宫廷的一次海洋探险于1935年8月19日采集到的。 76

## 被囊动物：难道人类曾是皮酒囊？

拉起海洋港口中系在浮标上的系泊绳索，你可能会发现绳索上面覆盖着数百个瓶状的皮质硬块，通常呈黄色或棕色，每一个都有几厘米长。从它们水下栖身的地方拉一只下来，它喷出的水可能会射入你的眼睛。尽管看起来很不像动物，但它们是动物，名为海鞘。这些块状的东西看上去很不成形，但它们是我们演化过程中的亲戚，和我们人类同属一个门：脊索动物门。从表面上看，海鞘被包裹在一层坚硬的外壳即被囊中，这层被囊摸起来更像是植物，而不是动物。很明显，这是因为被囊中包含纤维素，这种化学物质通常能在植物体内发现，但不存在于动物体内。海鞘的身体顶端有两根管子，确切地说是虹吸管，海水通过一根被吸入，通过另一根被排出，水流是依靠海鞘体内千百万根细小纤毛的摆动驱动的。持续不断的水流将细微的食物颗粒和溶解在水中的氧气输送到身体中，同时排出废物。

门是由一组演化上有亲缘关系、具有相似身体布局的动物组成的。再重复一下瓦伦丁的说法：“门是生命树上基于形态学的分支。”因此，海鞘怎么可能与你我一样，与鸟和鱼一样，同属脊椎动物呢？观察一个成年海鞘——一个常年静止不动的、滤食性的、披着纤维素外皮的块状物，几乎没有证据表明它与我们之间有密切的演化关系。实际上，早期的博物学家也完全没有意识到这种联系。亚里士多德认为海鞘属于软体动物，类似蚌和蜗牛，但他也指出了它们的与众不同之处，它们的“壳”，确切地说，是被囊，不是坚硬的，而更像是皮革，而且包裹住了整个身体。19世纪初，拉马克将它们从软体动物中移出来，为它们单独建立了一个新的类别——被囊动物，但他并没有解决它们的类缘关系问题。1866年才发生了重大变化，一位杰出的俄罗斯动物学家亚历山大·科瓦列夫斯基发表了一份关于海鞘胚胎发育和幼体发育的详细描述，同时他还意识到了自己所做出的发现的深远意义。海鞘的胚胎会发育成小小的“蝌蚪”，通常只有一毫米长，在海中游动生活一两天，然后头朝下地栖在岩石或其他的基底上。它们在那里会经历剧烈变形，成为小小的成体的样子。从那时候起，这种动物便再也不会移动位置，它们牢牢地待在栖身的位置，过滤海水。科瓦列夫斯基发现，在游动的蝌蚪阶段，它们的前端会长出一个小小的脑，连接着背部的神经索，而神经索正好位于一根坚硬的杆子的上方，这根杆子便是脊索。这些特征全都是脊椎动物的特征，人类和鱼类都有，至少是脊椎动物胚胎阶段的特征。被囊动物与脊椎动物在演化上的关系就此清楚明了了。

有关这个发现的新闻席卷了科学界，因为当时围绕着哪类无脊椎动物可能与脊椎动物是近亲有很多的争议。在1871年出版的《人类的由来》一书中，达尔文写道：

科瓦列夫斯基先生近期所做的一些观察，得到了库普弗教授的确认，将成为一项格外引人关注的发现……这项发现便是：海鞘类动物的幼体与脊椎动物是有联系的，表现在发育模式、神经系统的相对位置，以及与脊椎动物的背索近似的结构上。胚胎学一直被证明是分类学最安全的指南，如果我们以其为基础，那么显然，我们终于得到了脊椎动物起源的线索。

这种新出现的观点认为：演化出海鞘和脊椎动物的早已灭绝的共同祖先肯定是一种类似小蝌蚪的动物，具有能在如今的海鞘幼体上看到的各种特征。这一观点得到了达尔文的支持。但有几位其他的动物学家试图推论出脊椎动物是从更像现存的已经完成变形的成体海鞘的一个祖先演化而来的，这种观点一直持续到20世纪晚期。维多利亚时代的律师兼诗人查尔斯·尼夫斯曾经创作过很多关于演化论、啤酒和妇女权利的作品，他以打油诗的形式对后一种观点表示了支持：



世间奥妙知多少，个中缘由不明了！
有个问题想很久，为何世人爱喝酒。
结果冒出达尔文，揭开人类的真身：
解答完美又优良，原来人类曾是——皮酒囊！

尼夫斯可能为自己找到了尽情喝酒的理由（他在随后的八节中详细展开论述了这个理由），但事实上，他并没有准确地反映出达尔文的观点或科瓦列夫斯基的发现。没有必要假设海鞘和脊椎动物的共同祖先（我们的远祖）的生命周期与现代海鞘一样，在发育过程中要经历变形。事实上，海鞘现存的近亲幼形纲动物终生都不会经历变形，一直都以游动的蝌蚪的形式生存、进食和繁殖。

## 文昌鱼：生活在沙中的谜

脊索动物门可以分为三个演化组，或称亚门。除了被囊动物（如海鞘和幼形纲）和脊椎动物之外，还有一类很吸引人的海洋脊索动物，名为头索动物，通常被称为文昌鱼。它们身上的很多特征都是脊索动物的典型特征。它们有脑和沿着背部而非腹部的神经索（脊索）、身体每面都重复出现的肌肉，以及连接咽喉和外部世界的咽裂。这些特征正好构成一幅典型的脊索动物





的身体平面图。除咽裂之外的大部分特征都能在海鞘的蝌蚪幼虫阶段观察到，成体海鞘才具备这种过滤结构。而在鱼身上，能见到脊索动物的所有特征，鳃就是从咽裂发育而来的，但在发育过程中，鱼的脊索渐渐被骨头包围，甚至被挤压得不存在。而我们人类，在发育的某一个阶段，也拥有所有这些特征，但同样，脊索也只有在胚胎阶段才比较显眼，而我们的咽裂在胚胎阶段也只是凹槽，始终都没有形成真正的洞。但是，脊索动物特征最明显的动物是文昌鱼，即使是成体也都具备这些特征。它有人们所能期望的最清晰的脊索动物身体结构。

在全球各海域栖息地中共发现了约30个种类的文昌鱼，它们通常分布在热带或亚热带海域，但有时候会出现在冷一些的水域中。有一种生活在欧洲海岸，能够在地中海部分海域和英吉利海峡危险的埃迪斯通礁石及著名的埃迪斯通灯塔附近的水

域中发现，通常藏身在沙砾中。另一种常见于佛罗里达湾区附近的潮下沙中，还有一种在中国厦门市附近曾经数量丰富，那里甚至曾经一度是商业渔场，文昌鱼在捕捞后作为食物出售。所有种类的整体外观都和鱼差不多，体长只有几厘米，节段性重复的肌肉分布在突出的如同一根加强杆的脊索两侧。有弹性的脊索对抗着收缩的肌肉，这使得文昌鱼可以在需要时快速游动，比如当它们从沙子中冲出来将卵子或精子产在海水中时。文昌鱼的咽裂非常明显，用来过滤从口中吸入的海水中的藻类。和属于脊椎动物的真正的鱼类不一样，文昌鱼没有骨头，没有鳍从身体侧面探出，头部也远远没有那么复杂。文昌鱼具备了脊索动物基本的“底盘”，但还没有脊椎动物演化过程中出现的那些复杂情形。

一个世纪前，文昌鱼是整个动物学研究中最热门的话题。1911年，伟大的德国演化论生物学家恩斯特·海克尔写道，文昌鱼是“除人类以外所有动物中最重要和最有趣的”。我很愿意赞同他的观点。即便如此，对海克尔及其同时代的人来说，文昌鱼其实是提出了一道难题。很多动物学家认为文昌鱼是一种退化的脊椎动物，是一种丢失了很多典型特征的鱼。另一些人觉得这极为不可能，比如英国最伟大的比较解剖学家埃德温·斯蒂芬·古德里奇就认为这种想法“荒谬透顶”。古德里奇的观点建立在对文昌鱼的发育和解剖的仔细研究上，他认为文昌鱼保留了更原始的脊索动物组织，与早已灭绝的脊索动物祖先没有太大区别。这个观点最终得到了人们的认可，近年又获得了基因组序列的进一步证据。因此，和海鞘幼虫的位置相似，文昌鱼是无脊椎动物与脊椎动物之间另一个关键的联系。文昌鱼是

80

如今现存的具备脊椎动物大多数基本形态特征的动物。它确实有自己的一些特殊之处，尤其是它只有一只眼睛的头部，这让它很像希腊神话中的独眼巨人。这些微小的变化发生在自从文昌鱼和脊椎动物最后一次共享一个祖先以来的五亿年间；而正是在这同一个时间段中，鱼、两栖动物、爬行动物、鸟和哺乳动物出现并呈现出多样化发展。文昌鱼不是任何其他活着的动物的祖先，81 从所有脊椎动物早已灭绝的祖先到文昌鱼的变化似乎非常少。

第九章

# 后口动物总门（中）：脊椎动物的兴起

鼎盛时期的罗马人将鱼作为所有娱乐活动的主宰，有专门谱写的曲目来迎接鲟鱼、七鳃鳗和鲻鱼。

——艾萨克·沃尔顿，

《垂钓者大全》（1653）

## 大分化？

动物学的教科书中很常见的情况是，一些书只关注无脊椎动物，而另一些书则专注于脊椎动物。很多大学课程也按照这种划分来教授动物多样化课程。这种划分由来已久。让-巴蒂斯特·拉马克因其提出的后天获得的特征可以被遗传的不可信观点而被铭记至今，他在200年前就已经明确了无脊椎动物与脊椎动物之间的区别，写下了《无脊椎动物》一书。他并不是第一个画出这条界线之人，因为在2 000多年前，亚里士多德就将动物划分成了“多血动物”和“无血动物”，这在本质上与脊椎动物和无脊椎动物的划分是相同的。

尽管这种划分由来已久，广为流传，但很多动物学家已经指出其中存在一个根深蒂固的问题。动物中的大多数个体是无脊椎动物，已经被记录的动物物种中的大多数同样是无脊椎动物。无脊椎动物与脊椎动物在物种数量上的差异是巨大的，无脊椎动物有上百万种，而脊椎动物只有大约五万种。但问题远比单纯的数量不等要复杂。问题表现在动物的演化树上，表现在动物的生命史上。动物被划分成门，门代表的是演化树上包含相似身体结构的物种的分支。大约有33个动物门，而其中32个是纯粹的无脊椎动物。即便第33个，也不是一个完全由脊椎动物构成的门，而是同时包含了无脊椎动物和脊椎动物。这自然就是我们人类所处的门——脊索动物门，其中包含无脊椎的被囊动物、无脊椎的文昌鱼与脊椎动物。所有这些动物的身体结构存在着足够的相似性，因而会被分到一起。如果我们退后一步，再看动物界的多样化，我们会发现脊椎动物并不足够特别，甚至不足以自成一门。所以，这是否意味着脊椎动物不过是动物生命树上的一个小分支而已？

## 深层的不同

数量问题以及演化问题的确不容置疑，而与此同时，脊椎动物也有一些非常重要的特殊之处。事实上，从某些方面来说，它们是动物中的例外。最明显的一点是，地球上几乎所有"大型"动物都是脊椎动物。的确有一些大型的无脊椎动物，如枪乌贼、章鱼和巨大花潜金龟，但大多数无脊椎动物体长不足几厘米。而与之相比，在脊椎动物的世界中，大型动物则非常常见。鱼、两栖动物、爬行动物、鸟和哺乳动物，都有各自的巨人。12米长的鲸鲨、1.5米长的大鲵、30米长的恐龙（当然，已经灭绝了）、3米高的象鸟（很不幸也灭绝了）、30米长的蓝鲸，这些动物可能是纪录保持者，但它们都不过是一系列动物中最大的一个。事实上，在脊椎动物中，非常小的尺寸几乎是闻所未闻的。最小型的脊椎动物之一，是一种产于印度尼西亚的鱼——微鲤，其成体体长不足一厘米。即便如此，它和很多无脊椎动物相比，也是一个庞大的怪物了。



长得更大的一个关键在于用于输送氧气并移除体内深层活性组织中二氧化碳的复杂的静脉和动脉系统：高效的“闭合式血液系统”。顺便说一句，无脊椎动物中最大型的一些物种，如枪乌贼和章鱼，同样也有闭合的循环系统，不过这是独立演化出来的。另一个同样重要的特征便是脊椎动物赖以得名的基础：脊柱。动物界中，骨骼框架有很多种形式。很多蠕虫有基于液体的支撑系统，节肢动物有坚硬的外骨骼，棘皮动物有坚硬的碳酸钙内板。但脊椎动物的骨骼非常不同，非常卓著。某些脊椎动物的骨骼由软骨构成，这是一种由蛋白质构成的结实但又有弹性的组织，但大体上说，它就是骨骼。骨骼格外轻盈而坚韧，能极为高效地支撑庞大的身体，此外，它还有另一项出人意料的特性：它是能生长的。蛋白质和矿物质基质中混合有沉积为骨骼的细胞和消除骨骼的细胞。另外还有能感知机械压力的细胞，以及传递信息指示骨骼生长或收缩并根据变化的条件做出反应的细胞。骨骼一直都是动态的。骨骼是一种不同寻常的组织，是适合大型的、活跃的、生长中的动物的理想组织，无论它们生活在水中还是在陆地上。



脊椎动物拥有复杂的脑部和感觉器官，在这一点上与被囊动物和文昌鱼等亲缘关系最近的无脊椎近亲也不相同。从七鳃鳗到人类，所有脊椎动物的大脑组织结构都是非常一致的，同时也都拥有三种关键的感觉输入方式，分别是视觉（成对的眼睛）、化学（成对的嗅觉器官）和力学（察觉水中的压力变化或空气中的声音）。脊椎动物的整个头部是精妙而复杂的，以包裹大脑的颅骨为基础，其上的感觉器官都是朝向外部世界的。颅骨的胚胎发育过程展示出另一个奇特之处：一种名为神经嵴的特别类型的细胞。这些细胞产生于发育中的神经索的边缘，在胚胎组织中迁移，然后形成各种各样的结构，包括颅骨上的骨骼或软骨，以及下颌和鳃部的支撑。如果没有神经嵴细胞，脊椎动物无法构建复杂的、被保护起来的头部；如果没有神经嵴细胞，脊椎动物永远都无法成为主宰陆地和海洋生态系统的大型食肉动物和食草动物。

庞大的体形、高效的血液循环、动态生长的骨骼、复杂精妙的脑部、保护性的颅骨，以及复杂的感觉器官，这些特征组合起来，将脊椎动物与它们的近亲区别开来。它们与文昌鱼和被囊动物同属脊索动物门，但脊椎动物的身体要复杂精妙得多。脊索动物门内部的不同还要更加复杂。脊椎动物和无脊椎动物的基因组序列之间的对比揭示出一个极为有趣的事实。DNA序列清楚地表明，在脊椎动物演化的早期，在其最根基的阶段或此后不久，发生了一项重要变异。整个基因组——每一条单独的基因——都被复制了，然后又被复制了一次。脊索动物祖先拥有的每个基因，早期脊椎动物都有多达四个。一些“多余”的基因很快就消失了，但是多数基因没有消失，因此，脊椎动物在基因的多样性方面要远远超过大多数无脊椎动物。新的基因是否为新的脊椎动物特征的演化条件，这一点是存在争议的，然而无论如何，有一点是可以肯定的：无脊椎动物和脊椎动物的分化是不容被忽略的。

## 脊椎动物的演化树

脊椎动物的一种常见分类方式是分为鱼、两栖动物、爬行动物、鸟和哺乳动物。在很多方面，这样的分类是非常合理的，但是它并没有精确地反映出脊椎动物的系统发生树。其中一个问题是，鱼类物种繁多，并非全都沿着单独的一条演化线，和其他动物分开。被称作“鱼”的动物和很多其他的脊椎动物混杂在一起。“爬行动物”中也存在相同的问题，现存的爬行动物和鸟拥有同一条演化线。如果我们严格地根据演化史来给动物分类的话，那么名为鱼和爬行动物的类别就不该存在。

85

尽管情况错综复杂，我们根据化石、分子生物学和解剖学等途径获知的脊椎动物的演化之路却是简单明了的。最初演化的脊椎动物是鱼形的，但缺少咬合的颌部。它们的兴盛期在大约四亿年前，尽管当时出现了诸多物种，但这些没有颌部的神奇物种仅有两个家族幸存至今：七鳃鳗和盲鳗。有颌部的脊椎动物是从没有颌部的祖先演化而来的，这些早期的捕食者分出三条主要的演化线。这三个类别分别是软骨鱼纲（包括拥有软骨骨架的鲨鱼）、辐鳍亚纲和内鼻孔亚纲（又名肉鳍亚纲）。这三条演化线都包含水生的“鱼”，但肉鳍亚纲中还包括离开水进入陆地生活的脊椎动物。它们的肉鳍之内都有有力的骨架结构，它们后来演化为四足动物——有四条肢体的脊椎动物，其中包含两栖动物、“爬行动物”、鸟类（与某些爬行动物位于同一条演化分支上）和哺乳动物。

## 七鳃鳗和盲鳗：饕餮之味和黏液

由于没有能咬合的颌骨，七鳃鳗和盲鳗需要用其他方式来将食物摄入口中。成体的七鳃鳗有一个像吸盘的杯状物，其中包含一个圆形的口部，里面武装着一圈圈尖锐的牙齿，犹如锉刀

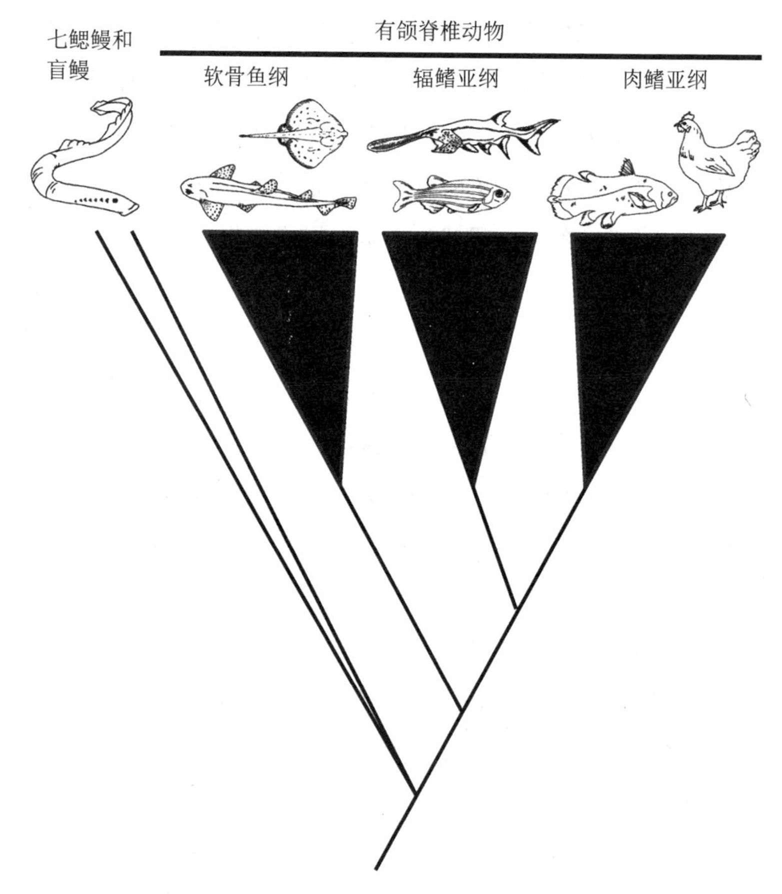


图12　脊椎动物的系统发生树

一般。这种外形骇人的器官令它们可以把自己牢牢地吸附在活着的猎物上吸食对方的血液，它们的猎物通常为大鱼。七鳃鳗可以固着在猎物身上好几个星期，一动不动地挂在上面，仿佛寄生的附属肢体。

七鳃鳗用吸盘附着在猎物身上，或者只是附着在河床中的石头上的时候，便没有办法通过口部从水中获取氧气。但七鳃鳗有“潮汐”鳃：水从它们头部侧面的孔洞被吸入，然后再从相 [86] 同的孔洞被排出。幼七鳃鳗不同于成体七腮鳗，它们只有普通的单向水流——从口部摄入，经过鳃部，然后从鳃裂中排出。幼七鳃鳗不是寄生的，所以可以依靠这样的呼吸方式生存。七鳃鳗在水位浅、河床多砾石的河流中产卵，孵化后，发育中的幼七鳃鳗会钻入厚厚的泥中，在其中居留数年，依靠从腐烂物质中提取的食物颗粒为食。在英国的很多溪涧和河流中，只要在水流较快的浅滩附近的泥沙中往深处挖掘，就能发现这些滑溜溜的蠕虫状的幼体。变形之后，吸盘形成，大多数成体会迁移到海中 [87] 生活。然而，有数个品种始终生活在内陆水域，比如欧洲溪七鳃鳗，它们的体长最多只有15厘米，相比于一米长的海七鳃鳗，真的是体形微小，而且它们不是寄生生活的。

七鳃鳗与君王们之间的关系由来已久。亨利一世国王，即征服者威廉的儿子，最爱的食物便是七鳃鳗。1135年，他在去诺曼底探访自己的孙辈期间，食用了“过量七鳃鳗”之后死亡。他的子孙后代却毫无畏惧，他的孙子亨利二世同样沉迷于这种没有颌部的美味，亨利三世要求定期为他提供烤七鳃鳗馅饼，“因为吃了七鳃鳗之后，所有的鱼都变得索然无味了”。1897年维多利亚女王登基六十周年庆典，1977年伊丽莎白女王登基二十五周年庆典时，格洛斯特市都送上了七鳃鳗馅饼，延续了这一王室传统。

盲鳗与七鳃鳗相似，也没有颌部，虽然没有吸盘，但围绕着两个侧向打开的咬合平板生有触须，同时还有一条角质化的、可伸缩的舌头。它们捕食海生蠕虫等活着的无脊椎动物，也搜刮海床上死掉的或濒死的鱼食用。盲鳗甚至会一路啃噬，钻入鲸或大型鱼类等体形较大的动物的尸体内，从内部蚕食它们。和七鳃鳗一样，盲鳗也缺少位于身体侧面的“成对的鳍”，成对的胸鳍和腹鳍是现存的有颌鱼类的典型特征，在陆生的后裔身上，则变成了腿。相比七鳃鳗，盲鳗的脊椎是发育不完全的，因此有些动物学家甚至不认为盲鳗是脊椎动物，而是使用有头动物这个词来概指盲鳗、七鳃鳗和有颌的脊椎动物。不过，这个观点是存在争议的，因为特征会在演化过程中消失，盲鳗的脊椎可能就遭遇了此类情形。我们不应该因为一种动物二次失去某些特征就将其从所属的自然群体中移走。就解剖学和胚胎发育模式整体而言，盲鳗与其他脊椎动物真的非常近似。



88

它们的确有一些特别之处，不过要说最特别的一点，非黏液莫属。很多动物也是分泌黏液的，但盲鳗将其发展到了一个新高度。盲鳗是毫无争议的黏液之王。盲鳗受到惊扰时，沿身体侧面排列分布的孔会释放出一种蛋白质分泌物，这种分泌物一接触到水，就立即大规模扩散开来，量大到惊人。一只20厘米长的小盲鳗，几秒钟就可以生成好几捧浓稠的像胶水一样的黏液，非常适合用于抵挡捕食者。为了避免被自己的黏液困住，盲鳗有一个很精明的小技巧。它会将自己扭成一个简单的反手结，然后沿着结滑动身体，将身体上的黏液蹭掉。

## 有颌：鲨鱼、鳐鱼、魟鱼

无论是否喜欢1975年上映的那部电影《大白鲨》，每个人都知道鲨鱼有颌部。颌部和成对的鳍是现存脊椎动物三大演化分类的决定性特征。这三个类别分别是软骨鱼纲（如鲨鱼和俗称狗鲨的点纹斑竹鲨）、辐鳍亚纲和内鼻孔亚纲。按照阿尔弗雷德·舍伍德·罗默的说法：“脊椎动物演化史上所有演化中最伟大的一个，可能就是颌的发育。”对狗鲨和其他有颌脊椎动物胚胎的研究已经清楚表明了这些高效的进食结构是如何演化的。在胚胎发育过程中，迁移的神经嵴流从形成中的后脑边缘向下移动，进入一系列的突起部位，形成鳃的骨骼支撑。从鲨鱼到人类，在各种有颌脊椎动物身上，其中的一个突起——颌弓，不仅发育成了鳃的支撑，而且发育为颌骨或颌软骨。位置在其后的细胞流，即舌弓，形成连接颌后部和颅骨的支撑结构。胚胎中细胞迁移的这些路径和模式表明，颌必然是从改良的鳃部支撑结构演化而来的。

现存鲨鱼的上颌并没有与颅骨融为一体，而是由有弹性的韧带非常松地悬挂在上面，并从后面辅以舌骨支撑。因此，鲨鱼 89 在进食时能够伸出两个颌，既可以从海床上精巧地捕捉小的猎物，也能将牙齿深深地咬入大型猎物的血肉中。大多数鲨鱼的牙齿都有锋利的锯齿，一旦将双颌深深地压入猎物体内，它们就能有效地切割对方的组织，同时鲨鱼左右摆动身体，进一步帮助切割。而寻找猎物的过程则涉及令人震惊而复杂的成套感觉器官。鲨鱼的嗅觉非常敏感，同时具有极强的定向性。有些鲨鱼， 90 特别是双髻鲨的两个鼻孔位于头部两侧，中间相隔的距离非常远，因而能更精确地推断出最高化学浓度的方向。在游向猎物附近的过程中，鲨鱼使用视觉感知和物理感知来探查水中的震动，作为导向的依据。而在发动攻击时，很多鲨鱼会在眼睛上罩上一个保护性隔膜以免受伤，这自然会令它们的视力即刻下降，但暂时性的目盲并不会让猎物获得逃生机会，因为这时的鲨鱼

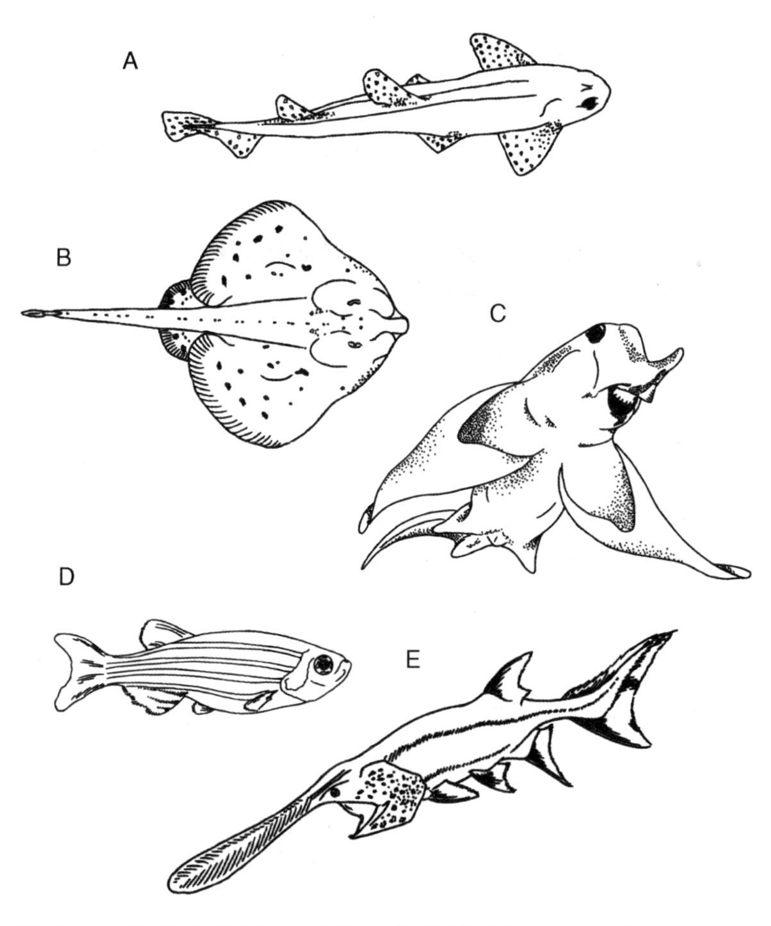


图13 A—C. 软骨鱼纲：A. 点纹斑竹鲨（狗鲨），B. 猬白鳐，C. 银鲛。D—E. 辐鳍亚纲：D. 斑马鱼，E. 匙吻鲟

依靠精妙的电感受器来探查动物肌肉产生的微弱电场。这些感觉细胞位于洛伦氏壶腹，这是鲨鱼皮肤上一系列特别的小洞。1678年，意大利解剖学家斯特凡诺·洛伦兹尼最早发现了这一器官，这位天才后来被托斯卡纳大公关入监牢，罪名是与大公分居中的妻子存在莫须有的友谊。

将鲨鱼、狗鲨、鳐鱼和魟鱼与大多数其他有颌脊椎动物区别开来的一个显著特征是，它们的骨骼是由软骨构成的，而非普通骨骼。它们体内还缺乏一个充满气体的腔体，这个名为鳔的结构能在辐鳍纲鱼类的身上发现，并演化成陆生脊椎动物的肺部。也许有人会认为，鲨鱼由于没有鳔，如果停止游动，就会沉到海底，但事实上并非如此。鲨鱼以一种截然不同的方式解决了浮力的问题。鲨鱼适应性地演化出一个巨大的肝脏，其中充满油脂，特别是长链碳氢化合物角鲨烯，其低密度抵消了鲨鱼骨骼、牙齿和鳞片的高密度，使得鲨鱼具有中性浮力。身体两侧结实的成对的鳍提供了稳定性和额外的升力。对于大多数鳐鱼和魟鱼等鲨鱼的近亲来说，浮力同样不是什么问题，因为这些动物通常是底栖的，也就是说，它们与海栖的自由游动的鲨鱼不同，是生活在海床上的。有些体形较大的魟鱼，如双吻前口蝠鲼，在海底停留的时间会比较少，它们在海中漫游，拍打十分巨大的胸鳍来移动，通过附着在鳃弓上的海绵组织形成的网过滤浮游生物。

软骨鱼纲中的最后一类，在演化上与鲨鱼、狗鲨、鳐鱼和魟鱼极为不同，这就是奇怪的银鲛。它们同样有软骨骨架，没有鳔，它们也像鲨鱼一样体内受精。但是银鲛与其他软骨鱼类不同的是，它们的上颌与颅骨连为一体，同时身体两侧均只有一个鳃，而非数个鳃。它们的头部结构坚实，但有一个像大象一样突出的肉质口鼻部，因而外形十分古怪。它们的外观大体上是鱼形的，鳍很大，它们的大眼睛和龅牙却非常像卡通兔子，有些品种还有一条长长的老鼠一样的尾巴。它们被俗称为奇美拉鱼，这自然令人联想到古希腊神话中由数种动物的身体部件混搭出来的怪物奇美拉，根据荷马在《伊利亚特》中的描述，它是一种"不朽的存在，是狮头羊身蛇尾的非人类"。

## 辐鳍鱼：柔韧灵活

大多数广为人知的鱼类品种都属于同一个包含丰富物种的类别：辐鳍鱼纲。其中的典型代表包括鳕鱼、黑线鳕、鲱鱼、金枪鱼和鳗鱼等商业鱼类，金鱼、脂鲤、孔雀花鳉和鲶鱼等养在水族箱中的物种，鳟鱼、鲤鱼、梭鱼、斜齿鳊和欧洲鲈鱼等垂钓者能钓到的大多数鱼类品种，此外还有很多品种，如米诺鱼、棘鱼、虾虎鱼等。相对而言，列出不属于辐鳍鱼纲的鱼要更简单一些，因为只有盲鳗、七鳃鳗、鲨鱼、鳐鱼、魟鱼、银鲛、腔棘鱼和肺鱼。全世界有超过24 000种辐鳍鱼生活在远洋、近海、河流和湖泊中。

和鲨鱼一样，辐鳍鱼既有成对的鳍（偶鳍），也有不成对的鳍（奇鳍）。不成对的鳍沿身体中线分布，包括背上的一个或多个背鳍、尾部的一个尾鳍，以及腹部的一个臀鳍。同时辐鳍鱼还有两组成对的鳍：位于鳃后的胸鳍，以及再后面一些的腹鳍。辐鳍鱼形如其名，它们的鳍由纤细的辐射线般的骨骼支撑，这种骨骼通常被称作鳍棘，鳍棘令它们具有极强的机动性。鳍棘对可以扭曲和活动的胸鳍来说尤其重要，无论鱼是在游泳、转身，还是在水中一动不动，胸鳍都能得到很好的控制。在辐鳍鱼的演化过程中，鳍经过了很多不同的改变，这显然是这一群体多样化的一个潜在原因。举几个极端的例子，裸背电鳗可以利用巨大的臀鳍产生的波纹缓慢地向前或向后游，而飞鱼有巨大的翼状胸鳍，可以在空中滑行长达50米。金枪鱼能以爆发速度冲向猎物，依靠的是将运动集中于尾鳍和身体后部；海马完全没有尾鳍，利用背鳍的波动来缓慢游动。

支撑鳍的鳍棘同样显示出了不同。某些物种的鳍棘具有防御功能，鲈鱼和棘鱼的背鳍上有尖锐而突出的骨刺，有几种鱼还能利用鳍棘注射毒液，如石鱼、龙䲢、蓑鲉等。有些鱼还会将鳍棘作为进食的辅助器官，如绿鳍鱼和鮟鱇鱼。绿鳍鱼是一种底栖类的鱼，胸鳍细长，上面有一系列感觉器官，可以用来在海床上“行走”，并感知猎物。鮟鱇鱼背鳍的前三根鳍棘异乎寻常地长，并连在一起，形成了一根钓竿，可以用来将猎物引诱到张着的嘴中。

辐鳍鱼以不同于鲨鱼的方式解决了密度问题。在它们的脊椎下面，有一个充满气体的腔体，这就是鳔，是它们体内的浮体，协助维持浮力。鲤鱼和鳟鱼等一些鱼的鳔由一根管道与肠道连接，因而它们能在水面上大口吸气来填充鳔。而鲈鱼等一些鱼的鳔与肠道的连接已经消失，它们的鳔依靠一个专门的腺体充气，这个腺体能吸收血液中的气体，再分泌到鳔中。很多淡水鱼类，包括鲤科的一些成员，还用鳔来增强听力，它们脊椎骨上改良的刺能将鳔的振动传递到内耳。说起来可能令人吃惊，有些鱼类还能利用鳔来发出声音，以吸引配偶或威慑竞争对手。比如雄性的蟾鱼（蟾鱼科）通过收紧鱼鳔上的“音肌”来发出声音，从而促使鳔的膜壁快速振动。由此发出的声音犹如响亮而悲戚的雾号角声。

辐鳍鱼的头部是复杂而精妙的。大多数辐鳍鱼的左下颌和右下颌能够向侧面打开，而上颌骨中的一块——前上颌骨，可以向前探出。这些运动使得口腔能突然间增大，创造出强劲的吸力，用于捕食猎物，否则这些猎物很容易逃掉。种类繁多的辐鳍鱼都采用这样的吸入式进食方式，这种进食方式支持着很多物种的生态平衡。头的后部是鳃，鳃都隐藏在名为鳃盖的盖片下面，鳃盖不仅保护着精巧脆弱的鳃，而且在鳃的功能中起着关键作用。合上鳃盖张大嘴巴，然后再闭上嘴巴开启鳃盖，辐鳍鱼便可以有效地驱动水通过鳃部，不游动的时候也可以完成。鳃丝总是处于与水流相反的方向，血液通过循环流向鳃丝，使辐鳍鱼能从水中摄取到最大量的氧气。



辐鳍鱼多样性的根源部分在于鳔的形成、鳍棘的适应性演化、吸入式进食和鳃盖。大量物种依靠的是很多因素的综合作用，如生态机会、适应性的身体结构，甚至可能还包括基因组的特征等。关于基因组的特征，很有意思的是，脊椎动物发生了两次基因组复制，而在辐鳍鱼中占据大多数的硬骨鱼在此基础上还多一次复制。我们还不能明确，这是否为身体结构更大的适应性提供了条件，或这是否导致了物种形成率的提高（通过在不同种群中丢失不同的基因）。多出的一次基因组复制并没有影响到所有的辐鳍鱼，这个大类下面现存的动物中还有一些是最早期的演化辐射的后裔：所谓的“无硬骨辐鳍鱼”，软骨硬鳞亚纲。其中包含头部像勺子、以滤食为生的匙吻鲟，身披重铠的雀鳝，以及各个种的鲟鱼，很多鲟鱼现在都十分稀少或濒临灭绝，它们的卵即为鱼子酱的原料。

第十章

# 后口动物总门（下）：陆生脊椎动物

蝾螈眼睛青蛙趾头，
蝙蝠绒毛恶犬舌头，
蝰蛇舌叉蛇蜥尖刺，
蜥蜴腿足枭鸟翅翼，
炼为毒蛊扰乱人世，
鬼神皆惊沸腾不止。

——威廉·莎士比亚，
《麦克白》第四幕第一场

## 从肉鳍到腿

1938年12月22日，有人给南非一位年轻的博物馆馆长展示了一条不同寻常的、彩虹般绚丽的、蓝色的鱼，这条鱼是由当地的一艘渔船捕获的，体长两米左右，有很强壮的肉鳍和重铠般的鳞片，被捕获后立刻引发了巨大关注。这是腔棘鱼的第一个活样本，腔棘鱼是一种古代鱼类，其化石记录可以追溯到四亿年前至6 500万年前，它们被认为已灭绝了6 500万年。《伦敦新闻画报》将这个发现描述为“20世纪自然史领域最惊人的事件之一”。那位博物馆馆长名为马里乔·考特尼-拉蒂默，于是这种鱼的学名便以他的姓氏命名为拉氏腔棘鱼，中文里一般称为矛尾鱼，其样本之后在非洲东海岸被多次捕获，特别是在科摩罗群岛附近，而在印度洋中发现了第二种腔棘鱼：印度尼西亚矛尾鱼。



发现依然存活的腔棘鱼之所以让人兴奋，并不仅仅是因为它们曾经被认为已经灭绝。更重要的是，它们对于理解陆生脊椎动物的演化具有重大意义，它们是我们人类演化史中的重要一步。肉鳍是争论的核心，它们可以在左右两侧独立移动，腔棘鱼看上去像在开阔的海洋中行走。肉鳍的结构和颅骨上的多处特征证明，腔棘鱼属于肉鳍类（肉鳍脊椎动物），而不是辐鳍鱼类。除了腔棘鱼外，还有两类现存的肉鳍脊椎动物——肺鱼和四足动物，后者中包括所有的陆生脊椎动物，人类也在其中。腔棘鱼和肺鱼都并非陆生脊椎动物的真正祖先，但这三类动物是近亲，都由大约四亿年前早泥盆世时期游动的肉鳍鱼类演化繁衍而来。化石证据和分子数据都表明，四足动物与肺鱼和腔棘鱼的关系都不亲近，但这两类肉鳍鱼对于理解我们自身的起源非常关键。现在依然存活的肺鱼中，四种在非洲，一种在南美洲，一种在澳大利亚，它们都是非常独特、非常不寻常的动物，但它们都确实是呼吸空气的鱼——它们的肺部和陆生脊椎动物的肺部相同。



无脊椎动物中的好几类，如昆虫、多足纲动物、蜘蛛和蜗牛，分别独立完成了从水生到陆生的困难变迁，而在脊椎动物的整个演化过程中，这种变迁只成功过一次。仅有的一条演化线上的脊椎动物克服了在陆地上生活的挑战，造就了今天仍生活在陆地上的所有脊椎动物：所有的两栖动物、所有的爬行动物、所有的鸟和所有的哺乳动物。要成功地生活在陆地上，动物必须能够从空气中获得氧气，在陆地上找到或捕获到食物，在支撑力比水相差甚远的环境中负荷自身重量，驱动身体在陆地上移动，避免由于过度缺水而脱水。肺鱼是陆生脊椎动物的近亲，它们既能用鳃在水中摄取氧气，也能用肺呼吸空气，这表明在陆地生活的真正变迁发生之前很久，空气呼吸就已经演化出来了。而在陆地上支撑身体、进食和移动，是更大的挑战，这些行为需要从鱼到四足动物的过程中发生多种演化改变才能实现。一些独特的化石揭示了这些变化，甚至揭示出了这些变化发生的顺序。

97



解剖结构上的变化之一是演化出能够猛地咬住猎物的扁平的口鼻部，而不再使用在水下运作良好的吸入式进食的方式。潘氏鱼和提塔利克鱼生活在约3亿7 500万年前，现已灭绝，它们的化石很清晰地呈现出这一特征，按化石推断，这些动物必然有一个类似鳄鱼的前端。不过它们的鳍依然有很明显的鱼类的特征，骨骼元素的末端是纤细的辐射束骨针，而不是强健的、长着骨头的指。由于这些既像鱼又像四足动物的特征的组合，提塔利克鱼的发现者尼尔·舒宾给这种动物起了个昵称：有脚鱼。生存时代稍晚的棘螈生活在3亿6 500万年前，鳍的末端是有关节的指状结构，这就使得它更像四足动物。有趣的是，和今天现存的大多数陆生脊椎动物不同，它们的前肢并不是仅有五指，而是有八指，后肢可能也是一样。几乎可以确定的是，棘螈生活在水中，用鳃呼吸，但它可能有能力到陆地上冒险，甚至在陆地上捕捉食物或晒太阳。另一种已经变为化石的早期四足动物鱼石螈可能代表了向陆地生活转变的另外一步，因为除了上述特征之外，这种动物还有着相对坚硬的中轴骨，也就是脊椎，上面有较长的骨质突起物，也就是脊椎上的关节突，它们将椎骨连接在一起，并帮助支撑动物的身体重量。

## 蛙和火蝾螈：皮肤呼吸

泥盆纪时代就已经灭绝的动物可能代表了脊椎动物向陆地最初的试探，但这些动物依然严重依赖水，特别是在繁殖过程中。直到今天，依然有一些四足动物是在水中繁殖的，它们在陆地上度过生命中的大部分时间，但在水中或水边产卵。这些动物包括蛙、蟾蜍、蝾螈、火蝾螈和没有腿的蚓螈，这些便是现存的两栖动物。大多数两栖动物始终都不会远离潮湿的栖息地，因为它们的皮肤防水性能不是很好，而且必须保持表面湿润来进行气体交换。依赖水生活的第二个原因是它们的卵和幼体需要潮湿的栖息地。大多数现存两栖动物的幼体（比如蛙的幼体蝌蚪）甚至有外鳃，能从水生环境中直接摄取氧气。我们在讨论两栖动物时，很容易会认为现存的两栖类动物是通往“真正”的陆地生活的一块普通而不重要的踏脚石，相比爬行动物、鸟和哺乳动物，它们没有那么成功，演化还不够，但这个想法是不准确的。它们存活至今这一事实，就足以证明它们持续的成功。实际上，现存的两栖动物非常特别，和最早期的陆生脊椎动物截然不同。其中一些品种数量庞大，尤其是几种蛙和蟾蜍。比如蔗蟾自1935年被有计划地引入澳大利亚，其后却造成了灾难性的影响，在澳大利亚北部扩散极广，数量巨大，已成为当地主要的入侵害虫。

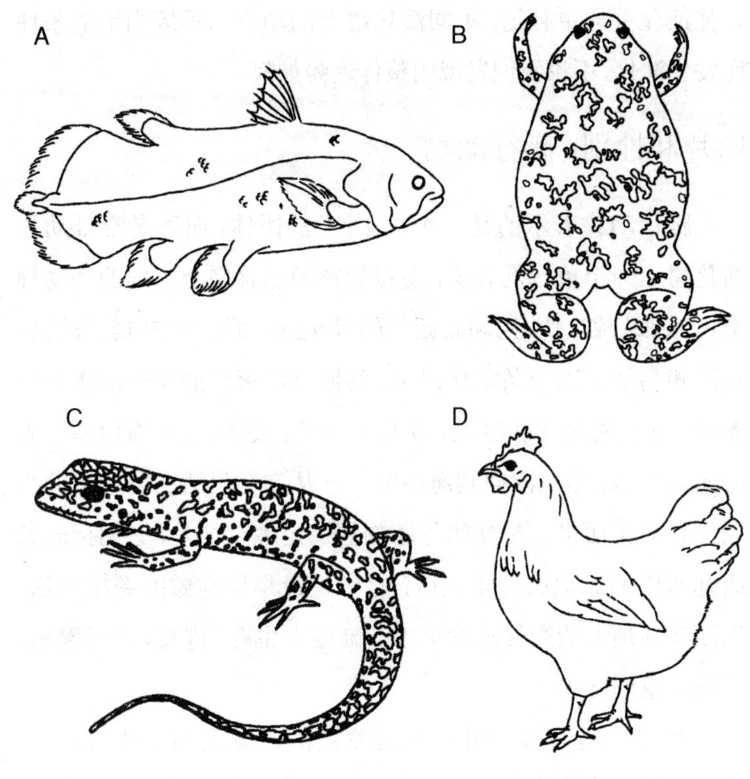


图 14　肉鳍动物：A. 腔棘鱼，B. 非洲爪蟾，C. 塔斯马尼亚雪蜥，D. 鸡



一些两栖动物终生都生活在水中，即便是成年之后，也从不试图到陆地上，其中包括能长到1.5米长的日本大鲵、长相古怪的隐鳃鲵属美国大鲵和非洲爪蟾。但“水生两栖动物”中最知名的可能是俗称墨西哥蝾螈的美西钝口螈，它的样子就像是一只20厘米长、性发育成熟的大蝌蚪，有着完整的羽状外鳃。事实正是如此，因为美西钝口螈是从“普通”的陆地蝾螈演化而来的，发育生理方面发生了变化，它们的发育成熟过程不需要经历变形为祖先的成体形式。美西钝口螈是强有力的提醒，让我们

记住演化并非单行道，不同演化线上的动物为适应当地的条件而发生演化，不论我们发现的整体趋势如何。

## 鳞片和性别：爬行动物

100

爬行动物表示的是一个“级别”的团体，而不是脊椎动物演化树上一个单一的分类，现存的物种包含完全不同的动物种类，如蜥蜴、蛇、龟、鳄和新西兰的喙头蜥。恐龙也是爬行动物，与鳄和鸟处于同一条演化线上，其他已经灭绝的爬行动物还有翼龙、海生的鱼龙和沧龙，以及蛇颈龙。这些水生物种和今天的海龟类似，是二次回到海中的——从完全生活在陆地上的祖先物种演化而来。早期爬行动物的决定性特征是，它们标志着离开水生栖息地的真正突破：它们是最早这样做的脊椎动物。生活在陆地上的爬行动物可以在陆地上生存、进食，甚至繁殖，无须再返回水中。



在这个转变过程中，两项关键的革新是处于核心位置的——演化出完全防水的皮肤和出现带壳并有数层内膜的卵。第一个特征十分容易理解，它的实现是由于演化出了更加复杂的皮肤，这样的皮肤包含数层能产生角蛋白和脂质的细胞。这个变化意味着皮肤不能再被用于呼吸（如现代的蛙和蝾螈那样），因为只有湿润的表面才能容许氧气和二氧化碳通过。爬行动物演化出了“肋式呼吸”，附着在肋骨上的肌肉被用于令肺部通气，将肺部转换成了更加高效的呼吸器官。而“有羊膜包裹的卵”的重要性看起来则没有那么明显，却至关重要，关键在于三层膜：羊膜、尿囊和绒毛膜。这三层膜将胚胎包裹起来，并提供了分布广泛的血管用于气体交换，同时还有一个储存有毒的含氮废物的地方，令发育中的身体能安全无虞。包括龟和鳄在内的大多数爬行动物产下带壳的有羊膜包裹的卵，不过有些蛇和蜥蜴是胎生的。束带蛇、巨蚺、蝰蛇等一般采用这种繁殖方式，这通常需要母体在体积不小的含有蛋黄的受精卵的整个发育过程中都将其保留在自己体内。还有些爬行动物直接由母体提供营养物质，而不是由蛋黄，在一些极端的案例中，甚至存在胎盘，如南蜥属和拟岛蜥属的石龙子。爬行动物很多生理上、解剖上和行为上的改变，令它们可以进入地球上最炎热、最干旱的环境中，包括非洲、大洋洲、亚洲和美洲干燥炙热的沙漠地区。





爬行动物的身体生理机制能很好地适应温暖的环境，因为大多数爬行动物会通过在太阳光线下暴晒来提升身体温度。因此，虽然它们的身体没有良好的隔热保温系统，它们也能维持高代谢率和活跃的生活方式。温度还以一种非常独特、格外不寻常的方式影响着很多爬行动物的生态：温度能决定它们后代的性别。如果美洲短吻鳄的卵在低于30摄氏度的条件下孵化，就会发育为雌性，如果卵处于33摄氏度的环境中，就会发育为雄性。这种现象被称作温度依赖型性别决定（简称TSD），与更常见的基因型性别决定不同。按照基因型性别决定模式，基因的不同控制着后代的性别，如哺乳动物的Y染色体上存在雄性决定基因。基因型性别决定模式看起来更加可靠，可为什么一些爬行动物（确切地说还有一些鱼类）会采用温度依赖型性别决定模式呢？环境中的变化如气候变化，可能促使温度依赖型性别决定模式的种群灭绝，因为所有的后代可能都会发育成相同的性别，这难道不是一个威胁吗？答案似乎在于对当地生态条件的适应性改变，伊多·佩恩、托比亚斯·乌勒及其同事进行的塔斯马尼亚雪蜥的研究就很好地证明了这一点。这种爬行动物的居住地区从零海拔地区到山区均有，值得注意的是，生活在低海拔地区的，采用温度依赖型性别决定模式，而高海拔地区的同种动物则采用基因型性别决定模式。造成这种差异的原因似乎是，位于低海拔地区的母亲们利用温度依赖型性别决定模式，在温暖的年份产下更多的女儿，这样它们就有最大的机会在较长的夏天中长得更大，有更强的生育能力，而在寒冷的年份，它们则转而产下更多的儿子，因为雄性在雪蜥当中相对不重要。高海拔地区则丧失了这种优势，那里的生长速度整体都慢，气温波动更加剧烈，温度依赖型性别决定模式若不让位于基因型性别决定模式，就可能会对性别比例造成严重破坏。

102

## 羽毛和飞行：鸟

恐龙是爬行动物中广为人知的一类，曾经在地球上的陆地生命中称霸长达数百万年的时间。最早的恐龙大约于2.3亿年前演化出现，然后分化为大小、外形和生活习性不同的许多物种，直到6 500万年前它们突然灭绝。或者更确切地说，它们表面上灭绝了。恐龙全部灭绝的流行观点是有些靠不住的，因为现今生存的一些动物是一类恐龙的直接演化后代，这类恐龙是兽脚亚目恐龙。已经灭绝的兽脚亚目恐龙中，很有名的种类有体形巨大的食肉恐龙暴龙和体形稍小但可能同样恐怖的伶盗龙，电影《侏罗纪公园》里曾经出现过这种恐龙，这令其广为人知。当然，暴龙和伶盗龙已经不复存在，但你每天都能看到它们的一些近亲。有一类兽脚亚目恐龙并没有在6 500万年前灭绝，而是在全球大灾难中幸存下来，分化繁衍直至今天。这便是鸟。

从演化的角度来看，鸟是一类没有灭绝的恐龙。鸟由恐龙演化而来这个观点最初由托马斯·亨利·赫胥黎于19世纪60年代提出。赫胥黎指出了兽脚亚目恐龙与一种已经灭绝的鸟（始祖鸟）在骨骼框架上的关键相似性，对始祖鸟的了解来源于一些保存完好的距今1.5亿年的化石。尽管始祖鸟有类似蜥蜴的特征，如牙齿和长长的骨质尾巴，但同时它长有翅膀和羽毛。因此始祖鸟被认为是最早演化出的鸟类之一，现在人们依然这么认为。赫胥黎的观点引起了争议，虽然所有生物学家都接受鸟由古代爬行动物演化而来这一观点，但它们是恐龙的直接后裔的观点很快不再流行。20世纪的大部分时间中，这个观点一直不受关注，直到20世纪70年代，经过耶鲁大学约翰·奥斯特姆的谨慎研究，它又一跃成为主流观点。但最激动人心、最决定性的证据直到20世纪90年代才出现——在中国发现了数组值得关注的"长羽毛的恐龙"的化石，这些毫无疑问是"不飞"的恐龙，但它们的身体和腿上覆盖着羽毛。长羽毛的恐龙不仅为鸟和恐龙之间的亲缘关系提供了强有力的证明，它们还显示出羽毛是早期的适应性改变，可能被用于保暖，但这为后来的飞行奠定了基础。





现代鸟类的羽毛有非常独特的结构。用于飞行的羽毛有精妙的非均匀分布的结构，能在下行过程中提供硬度和力量，同时也非常结实，并且轻盈得不可思议。它们有一根主轴，名为羽干，从羽干上伸出数量繁多的紧密排列的羽支，每一根羽支上又长着细小的有倒钩的羽纤支，彼此连接在一起。与此不同，用于身体保暖的下层羽毛绒羽并不以相同的方式彼此相连，因而能锁住空气，而不是产生一个平滑的表面。除了飞行和隔绝寒冷这两个基本功能外，羽毛还在防水、伪装和交流等方面发挥着作用。羽毛和飞行主宰着鸟类的整个生态系统及习性，共同影响着它们的演化。在飞行过程中，体重是个关键因素，因此，鸟便演化出非常纤细的空心骨骼，利用内部的支撑来加强骨骼。沉重的牙齿在演化中消失了，长长的尾巴也同样消失了。但比绝对重量更重要的是体重的分配，所以鸟类的解剖结构发生了变化，重心位置比大多数脊椎动物都要靠前，正好位于翅膀中间。后腿的“大腿”沿着身体两侧向前折叠，同时脚部拉长，促成了重心位置的移动；这一变化也解释了为什么鸟的膝盖似乎是朝后的——实际上，那并不是膝盖，而是踝关节。



今天存活的鸟有大约一万种，在所有的大陆、所有的海洋，都有鸟类生活和飞翔。其中有生活在南美洲的体形很小的蜂鸟，新几内亚丛林中独特的天堂鸟，能飞越安第斯山口的雄壮的美洲秃鹫，能在远离陆地数百英里的海浪上飞掠而过的剪嘴鸥，盘旋在草地边缘地区的红隼，神出鬼没的鹪鹩、知更鸟、画眉等。这么一描述，仿佛是一幅展示生物多样性的画面，但实际上，所有的鸟都非常相似——至少在解剖结构上。最惊人的例外应该是少数鸟类二次失去了飞行能力。企鹅独特的体形使其更加适应水中活动，而不再适应空中活动；鸵鸟体形巨大，不再能飞。这些例外提醒我们，飞行在解剖和生理上给鸟带来了巨大的限制。演化无法跳脱出物理法则。



## 奶和毛：哺乳动物

鸟类活跃的生活方式能够实现，完全是因为它们相对高的体温，高新陈代谢率加上羽毛提供的保温作用使得鸟类的体温维持在相对高的水平。另一类能生成并维持自身热量的陆生脊椎动物正是我们人类所属的类别：哺乳动物。不过，哺乳动物关键的保温材料是毛。和羽毛相比，毛的结构要简单很多，它由非常简单的α-角蛋白纤维束构成。尽管如此，层层叠叠的毛能够非常高效地锁住空气，保持温暖。由于这种保持热量的方式，哺乳动物得以在寒冷条件下外出并四处活动，这时太阳的光线还没来得及温暖它们的爬行动物近亲。与鸟类不同，哺乳动物不是从我们现在所熟悉的爬行动物演化而来的。在羊膜动物（有羊膜包裹的卵的陆生脊椎动物）的演化树上，一个分支上出现了蜥蜴、蛇、鳄鱼、恐龙和鸟，而邻近的一条演化线（如今已经灭绝的下孔亚纲动物）上则出现了哺乳动物。

除了有毛之外，所有哺乳动物都有的第二个重要特征是哺乳：产奶以哺育后代。这是一个决定性的、至关重要的适应性改变，它使哺乳动物在一年中的任何时候都可以繁殖后代，即便是在多种食物不易找到或随时间波动而变化的情况下。成年雌性可以在能够获取食物的时候储存食物，并以脂肪的形式储备能量，而后代则依靠从母亲那里吸食高能量的奶而得到营养供应。有经验的成年动物进行食物采集，极有可能会比幼年动物高效，这就意味着吃奶能使幼小的动物将更大比例的能量用于生长。

听来可能奇怪，吃奶也促进了哺乳动物复杂生态多样性的发展，这是通过它们复杂的牙齿演化实现的。理由如下：由于哺乳期的存在，新生的哺乳动物并不需要牙齿。这就意味着在牙齿长出前，颅骨和颌骨能得到充分生长，因此包括蜥蜴在内的大多数羊膜动物身上常见的牙齿连续替换的简单机制发生了变化。哺乳动物演化出了双套牙，也就是会产生两套牙齿：幼年期长出一套简单的牙齿，然后在几乎生长完全的颌骨上长出更加复杂的牙齿。因为牙的发育被推迟，所以哺乳动物的牙齿得以演化，变得能在上下颌之间精准匹配，这种特征名为“咬合”。有些动物的颌骨随着牙齿的增长而不断变大，它们难以实现咬合。咬合令哺乳动物拥有了咀嚼和研磨食物的重要能力，特别是坚硬的植物，它们能将肉从猎物的身体上彻底地撕咬下来。拥有这样厉害的器官，相比任何其他种类的脊椎动物，早期的哺乳动物能更多样化地拥有更多的食物来源和进食策略。

现存约有4 400种哺乳动物，数目不及鸟类的一半，然而在身体外观、体形和生活方式等方面，却有更丰富的多样性。其中只有五种单孔目类哺乳动物，也就是卵生的哺乳动物：鸭嘴兽和四种针鼹鼠。所有其他的哺乳动物都是真兽亚纲的，即胎生的哺乳动物。其中包含了数百种有袋类哺乳动物，它们生出非常不成熟的幼崽，然后将其装在育儿袋里养育，例如袋鼠、袋熊、负鼠、长鼻袋鼠、袋狸、树袋熊和袋獾。大多数现存的哺乳动物是有胎盘的，有更长的孕期，没有育儿袋。胎盘哺乳动物的生态多样性是令人震惊的，包括鼩鼱等食虫动物，羚羊、象、长颈鹿、野牛等食草动物，狐狸、狮子等狩猎的食肉动物，鼠、人等有什么吃什么的杂食动物，海牛等水生食草动物，海豹、海豚等水生食肉动物，甚至还有天空中的哺乳动物蝙蝠。



20世纪的大部分时间里，胎盘哺乳动物真正的演化史都是不明确的。这么复杂的情形中，谁与谁是关系最近的？这个问题现在已经快要解决了，尤其是在近年来应用了DNA序列技术之后。新兴的共识将胎盘哺乳动物划分为四大谱系。值得注意的是，这些谱系完美地映射到了已知的大陆地质史上，这表明胎盘哺乳动物的多样化随着世界主要大陆的分离而发生。其中一个谱系名为非洲兽总目，顾名思义，其中包含起源于非洲的哺乳动物的目，包括象、土豚和海牛。南北美洲的哺乳动物被命名为“贫齿总目”，包括食蚁兽、树懒、犰狳等。劳亚兽总目包含在北方超大陆劳亚古大陆上演化的广大动物，这片大陆包括欧洲的前身和亚洲大部分地区。劳亚兽总目中有猫、狗、鲸、蝙蝠、鼩鼱、牛、马，以及很多其他种类。最后一类哺乳动物名为灵长总目，包括鼠、兔，以及猴子、猿等灵长目动物。

退一步审视我们人类在动物演化树中的位置，会发现人类只代表一根小小的细枝。我们位于灵长目，而灵长目属于灵长 [107] 总目。灵长总目属于胎盘哺乳动物，胎盘哺乳动物属于真兽亚纲，真兽亚纲属于哺乳动物，哺乳动物属于羊膜动物，羊膜动物属于四足动物，四足动物属于肉鳍动物。肉鳍动物是三类有颌脊椎动物中的一类，属于脊椎动物，而脊椎动物属于脊索动物，脊索动物属于后口动物，后口动物属于两侧对称动物，两侧对称动物是动物演化大树上的一个分支。[108]

第十一章

# 魅力无穷的动物

这世上有已知的已知，有我们知道自己知道的事情。这世上有已知的未知，也就是说，有我们知道自己现在还不知道的事情。但这世上也有未知的未知，有我们不知道自己不知道的事情。

——唐纳德·H.拉姆斯菲尔德，

《美国国防部简报》(2002)

## 新的门，新的观点

动物学的历史一直都是一个观念不断变化的故事。围绕动物之间演化关系的辩论和争议持续了一个世纪，问题随着不断发现的新物种而变得更加复杂。每一年，我们对于数百个物种的解剖、生态、发育和行为，都有更进一步的认识。因此，我们非常应该退一步考虑问题：我们现阶段的知识到底有多准确？未来会大翻转吗？我们现在是否已经建立起了可信的框架，可以据此进一步深入研究动物生物学？我们必须首先问问我们是否真正了解了动物界多样化的全貌。

毋庸置疑，还有成千上万甚至几百万个动物物种依然未被发现。热带雨林和深海是两个充满了各种生命的生态系统，而对它们的科学探索却只触及皮毛。然而发现一个新物种，甚至是发现一千个新物种，并不会从根本上改变我们对动物生物学的理解。从其他角度说，发现物种自然是意义重大的；比如了解一个生态系统中的全部物种，有助于理解养分循环和能量流动的模式。这方面的洞见是非常重要的。但是大多数新发现的物种都是已知物种的近亲，所以，如果我们希望理解地球上动物多样性的整体模式，这些新物种的发现并不是关键。它们填充了细节，但不会促进我们的知识状态发生根本改变。

如果涉及更高分类层级的话，情况就不同了。动物分类学中最基础的分类是门。这里再次引用瓦伦丁的说法："门是生命树上基于形态学的分支。"因此发现一个新的门，的确会改变我们的知识状态。发现新的门，等于给动物演化树添加了新的分支。同样重要的是，这揭示出一种新的形态学：一种构建身体的新方式。这两方面（新分支和新形态学）的结合会改变我们关于某些独特的特征何时、为什么、以怎样的方式在演化中出现的观点：也许是对称、分节或中枢神经系统之类的基础特征。不过，还有未被发现的门吗？

本书中，我写到了33个不同的动物门，其中大多数已经为人所知很久了。20世纪后期，很多动物学家认为所有门都被发现了。然而令人意外的是，1983年，丹麦的动物学家莱因哈特·克里斯滕森发现了一个新物种，它与其他所有物种都截然不同，因而需要建立一个全新的门。他将这个门命名为铠甲动物门。这些动物体形很小，通常体长远远不足一毫米。它们附着在沙粒上，样子就像是微缩的茶壶或冰激凌筒。20世纪70年代，其他动物学家也注意到了这些动物，比如罗伯特·哈金斯，铠甲动物游动在水中的幼体根据他的名字被命名为哈金斯幼虫。但是最令人吃惊的是，这个新的门并不是在偏远的、人迹罕至的地方发现的，而是在法国的罗斯科夫沿海区域，一个繁忙的海洋生物研究中心就坐落在那里。

下一个发现的“新门”——环口动物门——同样也是过去被人忽略的。环口动物是一种生活在挪威海螯虾和螯龙虾口部的小小的共生动物。宿主十分常见，肯定有成千上万人吃下过离群的环口动物，但从来都不知道自己在消化怎样的神奇物种。值得注意的是，在1995年正式记录这种新动物的，还是克里斯滕森——他与彼得·芬奇合作，而最先观察到这种动物的人是汤姆·芬切尔。

2000年，第三种新的身体结构发布了，这次确实是来自一个偏远地区，很少有科学家到访过那里。克里斯滕森（又是他）带着一组学生去格陵兰岛附近的迪斯科岛做考察。在那里，学生们在一处冰冷的淡水泉水中发现了一些不同寻常的用显微镜才能观察到的动物。它们的体长只有0.1毫米到0.125毫米，口部探出结构复杂的颌，解剖结构与其他所有动物的都截然不同，至少值得被划定为一个新的类别，可能也该被确定为一个新的门。它们被命名为微颚动物。

那么，是否还有有待发现的门呢？极有可能。上述三个例子全都是很微小的动物，体长都远远不足一毫米，在微小的动物中，未来可能会有类似的发现。去“小型底栖生物”（生活在沙粒之中的动物）中寻找极有可能有所收获。在法国海岸发现铠甲动物表明，类似的发现可以出现在世界上任何一个地方，尽管如此，我得建议，遥远的深海栖息地可能是希望最大的地方。不过，如果你真的想发现一个新的动物门，我会建议你不必去寻找新物种。相反，新的门可能正位于旧的门之中。

## 来自旧门的新门？

听来可能有些矛盾，对已知动物门的很多改变（无论是发现了新的门，还是将旧有的门合并）之所以能实现，是由于对早先发现的物种进行了更详细的研究。一个门中包含来自同一条演化分支的动物，因此，如果新的数据表明，一个门中包含了表面上相似却来自演化树不同分支的物种，那么这个门就会被分成两个。没有别的办法。在过去20年中，这样的事情已经发生了数次，尤其是DNA序列数据被用于检测动物之间的演化关系之后。如果DNA数据显示出一个非常明显的古怪之处——一个物种在演化树上的位置不对，那么分类就不得不修改。包容旧物种的新门就此建立。

最重要但仍有争议的案例中涉及两类不同寻常的蠕虫：无腔动物和异涡动物。尽管这两类中都不包含广为人知的动物，甚至不包含常见的动物，但它们为科学界所知已经很长时间。因此，尽管这些物种并不是新发现的，但可能仍然值得为它们建立一个或两个新的门。无腔动物包括的是与扁虫相似的小型水生生物，通常体长几毫米。最容易找到的一个物种是薄荷酱虫，这种虫外观十分漂亮，呈亮绿色，因为有一种水藻生长在它们的体内。它们生活在欧洲沿海的沙滩上，特别是法国罗斯科夫附近的海岸线上，在湿润的水坑里常能发现它们形成的绿泥状“油层”。如果你缓缓地靠近这层泥，它们就会习惯性地因被惊扰而消失；这是由成千上万的绿色蠕虫组成的一层活的泥，在受到惊扰时，它们只是爬到了沙子之中。这些蠕虫以及很多与它们相似的蠕虫，传统上都被分在扁形动物门中，与“真正”的扁虫、吸虫和绦虫并列。一直以来，都有一些不同声音呼吁人们关注它们在解剖结构上的独特特征，不过，按主流观点来说，薄荷酱虫及其同伴一直属于扁形动物门。直到基因序列对比结果清楚地表明它们与扁虫、吸虫、绦虫根本不是近亲，一个新的门才被提了出来。

异涡动物的情形与此相似。这些动物比无腔动物大很多，最早在挪威一个峡湾中发现的物种博氏异涡虫体长好几厘米，近年在太平洋发现的几个物种还要更大一些。它们也是扁平的蠕虫，看起来并不怎么特别，就是非常简单的黄棕色蠕虫，除了盲端的肠道，几乎没有可以辨别的器官。它们也曾经被很多动物学家认为是扁形动物，但有一些科学家认为它们和棘皮动物或半索动物关系更近。一个以最初的DNA序列分析为基础的看法认为，异涡虫是一种软体动物，但这个结论很快就被证明是一个不幸的错误，当时从异涡虫身体里提取的DNA来自它的最后一餐饭，而非它自己的细胞。后来足够的真正的异涡虫DNA被提取出来，进行了很多的基因序列分析，结果表明，这种动物并非扁形动物，也不是软体动物、棘皮动物或半索动物，而是与其他动物类别都有明显区别的生物。2006年，一个新的门被提出。

我们可能会发现在已知的动物中还潜藏着更多的动物门，它们之前被错误地归类到动物系统发生树的错误部分。那么，该从何处开始研究呢？数十种很独特的无脊椎动物与它们理论上的近亲只有少量的共同特征。动物学家面临的挑战是，要确定其中哪些是这个门的畸变成员（演化改变了它们的身体结构），又有哪些几十年来一直误导着动物学家。比如，卡特丽内·沃索就将注意力投注在了一种不同寻常的水生蠕虫身上，这种蠕虫现在被认为是环节动物，但几乎没有普通环节动物的特征，甚至可能都没有分节。另有一种蠕虫既有环节动物的特征，也有扁形动物的特征。吸口虫是另一类存在疑问的动物，这是一种不同寻常的蠕虫，很像是环节动物，寄生在海百合身上。它们当中会有一个新的门吗？

113

另一种奇特的动物可能是地球上最奇怪的动物了，它们是水龙骨水母。这种小动物一生大部分时间都生活在鲟鱼卵中，鲟鱼卵被用于制作鱼子酱。它们从鲟鱼卵中出来，会分散成一大群小小的水母。它们可能的确与水母是近亲，是刺胞动物门的成员，如果真是那样，它们当然也是非常奇怪的一员。它们也可能和一种奇怪的蠕虫状寄生虫有亲缘关系，身体没有明确的前后上下左右，也没有中枢神经系统。这两种动物都拥有类似刺胞动物的刺胞囊结构。分子分析表明，它们的确属于刺胞动物门，这就意味着它们之前被归入的黏体动物门，实际上应该被纳入刺胞动物门。因此，新的数据不仅能生出新的门，同时也能将门从名单中移除。



## 未来的观点

这些不同寻常的动物是否能被分入它们真正属于的门，这个问题到底为什么重要呢？关键原因在于，每一次我们将一个独特的身体结构或独特的形态放在动物生命系统发生树上，我们关于演化路径的观点就会发生改变。以无腔动物和异涡动物为例，这两类动物都是两侧对称动物，但它们身体中线的位置缺少一条中央的中枢神经索。这自然与大多数两侧对称动物不同——蜕皮动物、冠轮动物和后口动物中的绝大多数都有一条主神经索。如果中枢神经索是两侧对称动物的一个普遍特征，这两个新的门可能是从动物演化树非常早期的分支上繁衍而来的吗？无腔动物和异涡动物，或者它们当中的一个，是在蜕皮动物、冠轮动物和后口动物分化之前（但在刺胞动物之后）就分出了分支吗？如果是这样，也许它们可以让我们惊鸿一瞥地看到在整合信息的主神经索演化之前，最初的两侧对称动物的身体是如何运作的。初步的分子分析表明，确实就是如此，至少无腔动物是如此，尽管结论还存在一些争议。另一项分子研究认为无腔动物和异涡动物属于后口动物，和棘皮动物、半索动物、脊索动物并列。如果这是正确的，它们为什么没有主神经索呢？它们是在演化中将神经系统遍布整个身体，从而失去了主神经索吗？或者，我们对于两侧对称动物的共同祖先的观点是错误的？这些都是有待解答的重要问题，它们的答案取决于无腔动物和异涡动物到底位于生命树的什么位置，尽管已经有大量的分子数据，但要确定这一点依然出奇地困难。

这种争议让我们不禁思考，我们是否应该对目前动物界的演化树有信心。“新系统发生树”指出，一些非两侧对称动物谱系（多孔动物门、扁盘动物门、栉水母动物门、刺胞动物门）在很早期就与发展为两侧对称动物的分支分开，而两侧对称动物后来又分成了三个大的总门：蜕皮动物总门、冠轮动物总门和后口动物总门。我们怎么能够确信这种设想是正确的呢？过去的这个世纪中，演化关系的假说发生了彻底改变，那么它们可能会再次改变吗？我预测不会。相反，我认为现在正应该对“新的动物系统发生树”抱有信心，至少对其主干部分。这棵系统发生树几乎完全基于所有动物基因的DNA序列比较。尽管最早的以分子为基础的树是根据一个或几个基因构建的，但随着大量的分析研究，每个物种都有超过100个基因被纳入分析，这棵树的基 115

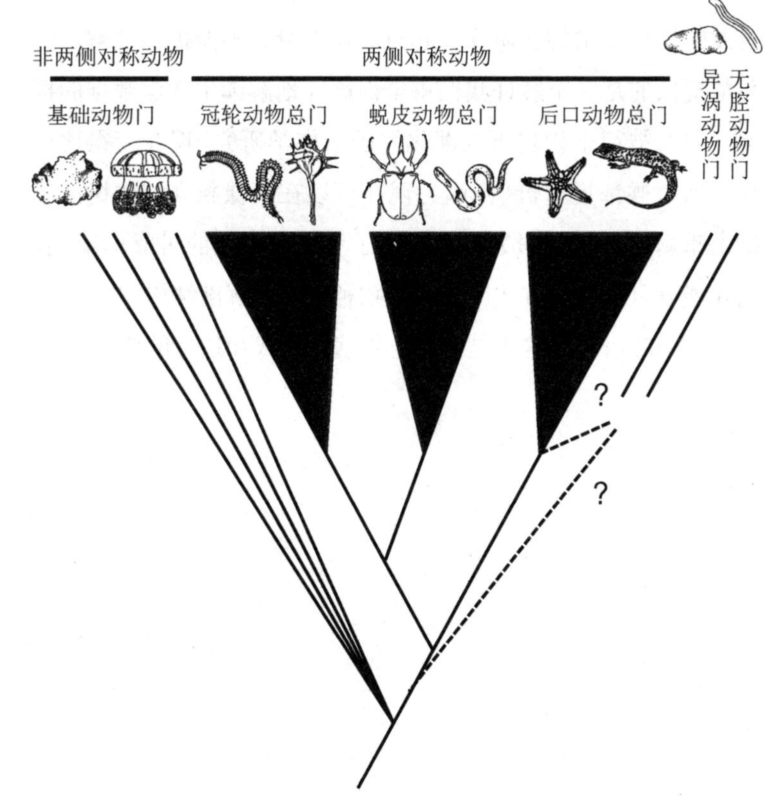


图15　展现另一种假设（关于异涡动物和无腔动物的位置）的动物界系统发生树

本框架已经被证实。DNA序列是关于过去历史的信息宝库，尽管分析起来并不简单，但它提供了关于这些问题的最可靠且内部最统一的数据组。的确，有些动物，如无腔动物，即便应用分子数据，也很难被定位，但至少，这些方法将它们区分了出来，或暂时放在存在争议的位置上，而不是随便挤在最方便的位置上。116

我相信，以动物学史的眼光来看，我们此时恰好第一次拥有了一棵可靠的动物多样性演化树。然而我们必须记住，这棵系统发生树只是生物学研究的起点。树本身并不能提供理解。它所能提供的是一个容许我们审慎而严谨地解读生物学数据的框架。曾经被用来构建不同演化树的形态学研究，现在变得比以往任何时候都更有价值，因为现在可以在一棵独立演化树的基础上重新解读这些研究。只有有了系统发生树的可靠框架，我们才能以有意义的方式比较动物物种之间的解剖结构、生理、行为、生态和发育，而这正是洞察生物演化模式和过程的路径。117

# 索 引

（条目后的数字为原书页码，
见本书边码）

## A



## B

## C



## D

## E



## F



## G



## H

## I



## J



## K



## L

## M



## N

## O



## P



## R

## S

## T



## U



## V



## W

Peter Holland

# THE ANIMAL KINGDOM

## A Very Short Introduction

# Contents

# Acknowledgements

The structure and content of this book owes much to past and present students at the University of Oxford and the University of Reading. Teaching a course in animal diversity to demanding and critical students has forced me to think carefully about the subject; student feedback has also helped highlight key issues. I acknowledge assistance from Merton College, Oxford, and members of the Department of Zoology, University of Oxford, notably Simon Ellis and Penny Schenk. I also thank Max Telford, Claus Nielsen, Bill McGinnis, Stu West, Theresa Burt de Perera, Tobias Uller, Sally Leys, and Per Ahlberg for comments on various sections, and Tatiana Solovieva for drawing the diagrams.

# List of illustrations

# Chapter 1
# What is an animal?

> I am the very model of a modern Major-General,
> I've information vegetable, animal, and mineral.
>
> Gilbert and Sullivan, *The Pirates of Penzance* (1879)

## To build an animal

In our everyday experiences, it is simple to decide which living things are animals and which are not. Walking through a town we may encounter cats, dogs, birds, snails, and butterflies, and we recognize all of these as animals. We should also include humans in our list. In contrast, we would have no doubt that the trees, grasses, flowers, and fungi we encounter are not animals, even though they too are living organisms. The problem of defining or recognizing an 'animal' starts to arise when we consider some of the more unusual living organisms, many of which are microscopic. It is helpful, therefore, to search for precise criteria for answering the question 'What is an animal?'

One feature shared by all animals is that they are 'multicellular'. That is, their bodies are made of many specialized cells. By this criterion, single-celled organisms such as the familiar *Amoeba* are not considered to be animals, contrary to the views of a century ago. Indeed, many biologists now carefully avoid the term 'protozoa' for organisms such as *Amoeba*, since by definition an organism cannot be both 'proto' (meaning 'first' and implying one cell) and 'zoa' (implying animal).

Having a body built from many cells is a necessary criterion, but it is not sufficient on its own. The same property is also found in plants, fungi, and some other organisms such as slime moulds, none of which are animals. A second important character of animals is that they get the energy necessary for life by eating other organisms, or parts of organisms, either dead or alive. This is in contrast to green plants, which can harness the Sun's energy using the chemical reaction photosynthesis taking place inside chloroplasts. There are plants that supplement photosynthesis with feeding (for example, the Venus fly trap) and animals with living green algae inside them (for example, corals and green hydra), but these do not blur the essential distinction too much.



Another feature that is often cited is the ability of animals to move and to sense their environment. This criterion holds up well for animals, but we need to remember that many plants have parts that can move, while cellular slime moulds (which are not animals) can form a slowly migrating slug-like structure.

The generation of sperm and egg cells of quite different sizes is another property typical of animals, and one with profound implications for the evolution of animal behaviour, but it is not a character that is readily observed. Perhaps the most consistent structural character is to be found when the cells of adult animals are examined closely. Although animals have many different types of cells, there is one type that has influenced the entire biology of animals and the evolution of the Animal Kingdom. The cell in question is the epithelial cell. These are brick-shaped or column-shaped cells, lacking the rigid cell walls found in plants. Epithelial cells are arranged into flexible sheets with specialized proteins holding neighbouring cells together and other proteins sealing the gaps between cells to make a waterproof layer. Sheets of cells are also found in plants, but their structure is quite different, being less flexible and more permeable.

The epithelial cell sheet of animals is remarkable for both functional and structural reasons. Epithelia can control the chemical compositions of liquids either side of the sheet, allowing animals to create fluid-filled spaces for purposes as diverse as bodily support or the concentration of waste products. Fluid-filled spaces were among the earliest skeletal structures of animals and were a factor permitting an increase in size during evolution, together with energy-efficient locomotion.

In addition, epithelial cell sheets are strong but flexible, supported by a thick layer of proteins such as collagen, allowing precise folding movements to occur. This is particularly important during embryonic development in animals, when folding movements are used to generate the structure of the animal body, rather like miniature origami. In fact, it is quite simple to mimic the earliest stages of animal development using sheets of paper. Although the details differ between species, typical animal development passes through a stage comprised of a ball of epithelial cells (the blastula), which itself was formed by a series of cell divisions from a single cell – the fertilized egg. In most animal embryos, the ball of cells then folds inwards at one point or along a groove, moving some of the cells inside. This event, which forms a tube destined to become the gut, is the crucial step called gastrulation. The indented ball is called a gastrula. Further folding events can occur to form fluid-filled supporting structures, muscle blocks, and – in vertebrates such as ourselves – even the spinal cord and brain. In short, cell sheets build animals.

All these characters are criteria by which we recognize animals, and they give insight into the basic biology of animals. But they do not comprise the most precise definition of an animal. In taxonomy, the classification of living organisms, names are given to branches – large or small – on evolutionary trees. An essential rule is that real or 'natural' groupings must encompass sets of organisms that have a shared evolutionary ancestor. This means that the term 'animal' must refer to a group of related species. The word cannot be applied to living organisms from elsewhere in the evolutionary tree, even if they possess some animal-like characters. Likewise, we would still use the term 'animal' for species that had lost some of the normal animal characters that were present in their ancestors. For example, some animals have lost distinct sperm and egg cells in evolution, while others are not clearly multicellular in every part of their life cycle, but since they share an ancestor with other animals, they are defined as animals. The animals, therefore, are a natural group (or clade) descended from a shared common ancestor. This clade is called the Animal Kingdom, or Metazoa.

## The origin of animals



From what did the long-extinct ancestor of all animals evolve? This sounds a difficult problem to solve, since the ancestor in question has been extinct for perhaps 600 million years, was certainly microscopic, and has left no fossil record. Surprisingly, the answer is known with considerable confidence. Furthermore, it was first suggested over 140 years ago. In 1866, the American microscopist, philosopher, and biologist Henry James Clark noted that the feeding cells of sponges, which are certainly animals, look quite similar to a little-known group of single-celled aquatic organisms known then as the 'infusoria flagellata'. Today, we call these microscopic organisms the choanoflagellates (collared flagellates) and DNA sequence comparisons confirm that they are indeed the closest relatives of all animals. Choanoflagellates and the feeding cells of sponges each have a circle, or 'collar', of fine tentacles at one end, such that they resemble miniature badminton shuttlecocks, plus a single long flagellum (a moveable whip-like structure) emerging from the middle of the collar. In choanoflagellates, the wafting or beating of the flagellum sets up water currents to bring food particles towards the cell, where they

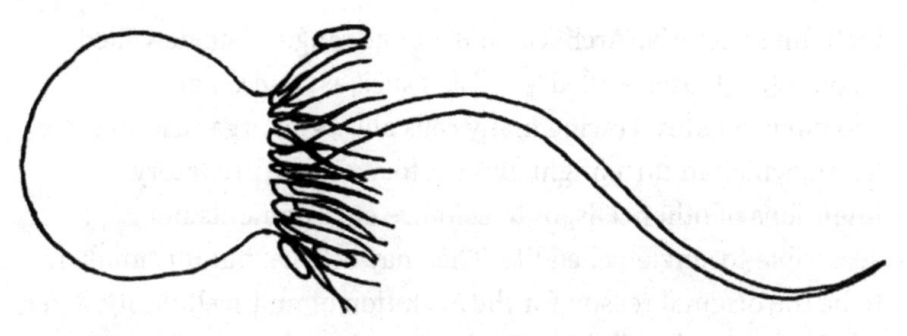

1. **A choanoflagellate, *Monosiga brevicollis*, feeding on bacteria**

are trapped by the collar. The feeding cells of sponges operate in a different way, but they still use the flagellum to create a water current. The most recent ancestor of all animals, therefore, was probably a microscopic ball of cells, each cell having a flagellum. The origin of the Animal Kingdom involved a series of changes that caused a shift from life as a single cell to life as a miniature aquatic ball of cells.



Recall that animals are not the only multicellular organisms on Earth: plants, fungi, and slime moulds are other examples of life forms built from a multitude of cells. These groups, however, did not arise from the same ancestor. Each evolved from a different single-celled organism. The multicellular plants are not closely related to animals or choanoflagellates, and they evolved in a completely different part of the tree of life. Fungi, such as mushrooms, brewer's yeast, and athlete's foot, are nowhere near the plants, and they again evolved from their own single-celled ancestor. Perhaps surprisingly, the fungi and their ancestors fit into the same part of the tree of life as the animals and the choanoflagellates, a group called the Opisthokonta. Multicellularity evolved twice in the opisthokonts: once to give animals, and once to give fungi. It is a sobering thought that we are more closely related to mushrooms, than mushrooms are to plants.

Why should multicellularity evolve at all? After all, the vast majority of individual living things on Earth have just a single cell, including bacteria, 'Archaea', and a huge range of single-celled eukaryotes (loosely called 'protists') such as *Amoeba* and choanoflagellates. Having many cells allows an organism to grow larger, which in turn might allow it to avoid the predatory intentions of other cells, or to colonize environments not accessible to single-celled life. This may be true, but it is unlikely to be the original reason for the evolution of multicellularity. After all, the first multicellular organisms, such as the ancestor of the animals, was probably little more than a microscopic ball of cells, confined to much the same habitat and way of life as its single-celled collared flagellate relatives. The problem of the origin of animals remains unsolved, although some intriguing ideas have been put forward. One clever suggestion, proposed by Lynn Margulis, is that multicellularity allows a fundamental division of labour: some cells can divide and grow, while others carry on feeding. But why should a single cell not divide and feed at the same time? The idea is that in single-celled organisms with a flagellum, such as the collared flagellates, a key part of the cell's machinery (known as the 'microtubule-organizing centre') must be used either for moving chromosomes during cell division or for feeding by anchoring the waving flagellum, but not both at the same time.

Another imaginative model has a slightly macabre appeal. The crux of the idea, proposed by Michel Kerszberg and Lewis Wolpert, is that self-cannibalism was an initial driver for the evolution of multicellularity. Consider a population of single-celled organisms, such as collared flagellates or their relatives, together with a few mutants in which the daughter cells do not fully separate after cell division. These mutants will form clumps or colonies of cells. Both forms feed on bacteria which they filter from the surrounding water. When food is abundant, the single-celled organisms and their colony-forming sisters can all acquire nutrients and reproduce successfully. At other times, however, food may become scarce, perhaps due to a change in the environment. Inevitably, many organisms will die as they fail to obtain enough nutrients to maintain basic cellular processes. The colony-forming mutants, however, have an instant failsafe mechanism for times of hardship: cells can eat their neighbours. This could happen by sharing of nutrients, allowing just a proportion of cells in the colony to survive, or more dramatically, by some cells decaying into food for adjacent cells. The consequence is that the colony-forming mutation would have a selective advantage in times of food shortage, and more of them would survive to reproduce. Self-digestion may sound grim, but it is actually a strategy used by several animals, from flatworms to humans, in times of starvation.

The whole of the Animal Kingdom has its origins in these ancient colonies of cells. Over the past 600 million years, possibly longer, the descendants of these cell colonies diversified and radiated through evolution, giving rise to the millions of different animal species on Earth today. Animals originated in the sea, but they have since colonized fresh water, land, and air. Examples include the flatworms and fishes found in streams and rivers, snails and snakes on dry land, and butterflies and birds in the air. Some, such as flukes and tapeworms, have invaded the bodies of other animals, while a few, such as dolphins, have returned to sea again. This great diversification spans a huge size range. The parasitic myxozoans and dicyemids have shrunk and simplified so that they are no larger than tiny colonies of cells, while giant whales steer their 100-tonne bodies gracefully through the oceans. To make sense of this vast diversity, we need to focus on the most fundamental unit of classification in the Animal Kingdom: the phylum.

# Chapter 2
# Animal phyla

> Classifications are theories about the basis of natural order, not dull catalogues compiled only to avoid chaos.
>
> Stephen Jay Gould, *Wonderful Life* (1989)

## Patterns and branches

For centuries, naturalists and philosophers have struggled to make sense of the range of life on Earth. One of the earliest and most pervasive ideas was that of a 'Scale of Nature' in which living, and sometimes non-living, things were arranged into a linear hierarchy. Each ascending rung on a ladder represented increasing 'advancement', based on a blend of anatomical complexity, religious significance, and practical usefulness. The idea had its origins in the thinking of Plato and Aristotle, but was crystallized by the work of the 18th-century Swiss naturalist Charles Bonnet. In Bonnet's scheme, the Scale of Nature rose from earth and metals, to stones and salts, and stepwise through fungi, plants, sea anemones, worms, insects, snails, reptiles, water serpents, fish, birds, and finally mammals, with man sitting comfortably on top. Or almost on top, being marginally trumped by angels and archangels. It is easy to ridicule such ideas today, but Bonnet had a good knowledge of the natural world. For example, it was Bonnet who discovered asexual reproduction in aphids and the way that butterflies and their caterpillars breathe. Furthermore, the idea of a Scale of Nature still pervades much modern writing, with many scientists talking of 'higher' or 'lower' animals: language that bears an uncanny resemblance to this old and discredited idea.

The dismantling of the Scale of Nature occurred gradually. A significant blow came from the respected French anatomist, palaeontologist, and advisor to Napoleon, Baron Cuvier. From his detailed studies on the internal anatomy of animals, Cuvier reached the conclusion that there were four fundamentally different ways to construct the body. These were not superficial differences, but were deeply rooted in the organization and function of the nervous system, brain, and blood vessels. In 1812, Cuvier organized the Animal Kingdom into four great branches (*embranchements*), named the Radiata (circular animals such as jellyfish, plus, surprisingly to modern biologists, starfish), Articulata (animals with a body divided into segments, such as insects and earthworms), Mollusca (animals with a shell and a brain), and Vertebrata (with bony skeletons, muscular heart, and red blood). No system was proposed to link these *embranchements*, and hence they stood parallel to each other, with equal status rather than as a hierarchy.

Cuvier, unlike his contemporary Lamarck, was not a believer in evolution. Paradoxically, however, it is evolution that provides the logical reason why Cuvier's *embranchements* could stand with equal status. As argued later by both Charles Darwin and Alfred Russel Wallace, evolution explains why every animal species has similarities with others, and why groups of species with common features can be identified. To exploit the familiar simile of evolution as a branching tree, or Wallace's more poetic phrase a 'gnarled oak', we should be able to describe small 'twigs' of closely related species embedded within larger and larger 'branches' including yet more distant relatives, all sharing common ancestry in evolution. We can then offer meaningful names to the small and large branches of the tree. The large branches within the Animal Kingdom are 'phyla' (the singular being 'phylum').

The tree analogy highlights the crux of the classification system for animals: names must reflect the natural relationships generated by evolution. Naming groups of animals is quite different from classifying a collection of inanimate objects, such as teapots, postage stamps, or beer mats. Inanimate objects can be grouped into multiple alternative arrangements, based on different properties such as colour, size, or country of origin, all of equal relevance. To classify living organisms in such a way would be to miss the fundamental point: a classification system based on evolution reflects natural order. It is a statement of relatedness, a hypothesis that proposes a particular evolutionary history.



## The list of life

How many animal phyla are there? In other words, how many 'large' branches of the evolutionary tree of animals are there? This begs the immediate question of how large (or small) a branch must be to warrant being called a phylum. This is a controversial issue, but in practice animals within the same phylum should share particular anatomical structures or features different from other phyla. In James Valentine's words, 'phyla are morphologically-based branches of the tree of life'. Phylum names can never be used to group together animals from different branches, nor should one phylum be nested within another. These rules work very well for most of the Animal Kingdom, and for all of the very familiar types of animals, but there are still disputes and disagreements concerning how many phyla are needed to classify the less well-known species. One thing for certain is that Cuvier's four categories represent far too gross a simplification; the number of animal phyla most often quoted today is between 30 and 35.

In recent years, several 'new' phyla have been proposed. This sometimes happens when a phylum needs to be split in two because research has found that it mistakenly contains animals from distinct branches of the animal tree. One example is the splitting of the former phylum Mesozoa into two phyla: the phylum Rhombozoa, containing some tiny worm-like parasites, and the phylum Orthonectida, containing more tiny worm-like parasites, these ones living rather improbably in the urine of octopus and squid. A more controversial example concerns the phylum Platyhelminthes (flatworms, tapeworms, and flukes), out of which some species have recently been removed and placed in a new phylum, the Acoelomorpha. New phyla are also proposed if totally new species are found with unusual and apparently unique body structures, and which do not fall within a pre-existing phylum. Both criteria must be satisfied in order to erect a new phylum. Since the 1980s, this has occurred just a few times, notably with the discovery of the phylum Cycliophora (tiny animals living attached to the mouthparts of lobsters and scampi), the phylum Loricifera (miniature urn-shaped animals found clinging to sand grains), and the phylum Micrognathozoa (even smaller animals found in a fresh-water spring in Greenland).



Phyla also get lost. This is not through extinction, or at least we can say that no phylum has gone extinct since human records began. Instead, phyla cease to be valid when it is discovered that the whole group actually fits inside another phylum. The two groups must, logically, be merged into one. This happens surprisingly often, most usually when a group of animals with very strange anatomy was originally classified as a distinct phylum, only for later research to reveal that they are actually greatly modified members of another group of animals. The best-known example concerns the giant tube worms, or pogonophorans, which are famous inhabitants of deep-sea hydrothermal vents around the Galapagos Islands and the mid-Atlantic ridge. Considering that some pogonophorans grow to 2 metres in length, it is surprising that their evolutionary relationships proved hard to track down. However, DNA sequence data now indicate that pogonophorans are modified members of the phylum Annelida, which contains such well-known animals as earthworms and leeches. Another example concerns the former phylum Pentastomida, or tongue worms, containing large (up to 15-centimetre) scaly parasites that hook inside the nasal passages of birds and reptiles. Despite their horrific appearance, it is now clear from DNA and cell structure analyses that tongue worms are actually highly modified crustaceans, placed well within the phylum Arthropoda and close to fish lice.

Here, I recognize 33 animal phyla. Of these, 9 phyla contain well-known animals familiar to almost everybody. A further 4 phyla can be found with a little effort, in ponds, in gutters, or by a walk along the sea shore. The 9 very familiar animal phyla are: Porifera (sponges), Cnidaria (jellyfish, corals, and sea anemones), Arthropoda (including insects, spiders, crabs, and centipedes), Nematoda (roundworms such as the river blindness parasite of humans, and the slug-killer nematodes used by gardeners), Annelida (earthworms, ragworms, and leeches), Mollusca (including snails, oysters, and octopus), Platyhelminthes (flatworms, flukes, and tapeworms), Echinodermata (starfish and sea urchins), and Chordata (including fish, frogs, lizards, birds, and mammals such as humans). The 4 additional phyla to be found with relative ease are Bryozoa ('moss animals', easily seen as arrays of tiny brick-shaped boxes on the fronds of seaweed), Nemertea (squidgy, slow-moving 'ribbon worms' found under rocks on the sea shore), Rotifera ('wheel animals' present by the thousand in pond water), and Tardigrada (miniature 'water bears' found in moss and topping the cuteness list for most zoologists).



In attempts to understand how animals have diversified through evolution, how they function, and how they are adapted to particular environments, it is best to start at the level of the phylum. Since phyla are 'morphologically-based branches of the tree of life', it follows that knowing which phylum a species belongs to helps us when making comparisons to other members of the same, or a related, group, and to consider how the anatomy of that organism relates to function. For example, knowing that an animal is in the phylum Nematoda draws immediate attention to the thick elastic cuticle and pumping pharynx found in this phylum, which is relevant to understanding the lifestyle and properties of that animal. Conversely, ignoring classification leads to confusing comparisons between distantly related species, which may have very different body plans and different constraints on their evolution and way of life. But we must not consider the animal phyla simply as a list of 33 different categories. They each comprise a branch of the evolutionary tree. And, of course, branches are always connected to other branches. Because of this, some phyla are more closely related to each other than they are to others. This information is vital for understanding how structure, function, and evolution interconnect in the Animal Kingdom.



Table 1

| **Phylum** | **Where in tree?** | **Examples** |
|---|---|---|
| Placozoa | Basal animals | |
| Porifera | Basal animals | Sponges |
| Cnidaria | Basal animals | Jellyfish, coral, sea anemones |
| Ctenophora | Basal animals | Comb jellies |
| Annelida | Lophotrochozoa | Earthworms, ragworms, leeches |
| Mollusca | Lophotrochozoa | Snails, oyster, squid, octopus |
| Nemertea | Lophotrochozoa | Ribbon worms |
| Brachiopoda | Lophotrochozoa | Lamp shells |
| Phoronida | Lophotrochozoa | Horseshoe worms |
| Bryozoa | Lophotrochozoa | Moss animals |
| Entoprocta | Lophotrochozoa | |

| | | |
|---|---|---|
| Platyhelminthes | Lophotrochozoa | Flatworms, flukes, tapeworms |
| Dicyemida | Lophotrochozoa | |
| Rotifera | Lophotrochozoa | Wheel animals |
| Gastrotricha | Lophotrochozoa | |
| Gnathostomulida | Lophotrochozoa | |
| Micrognathozoa | Lophotrochozoa | |
| Cycliophora | Lophotrochozoa | |
| Arthropoda | Ecdysozoa | Insects, spiders, crabs, centipedes |
| Onychophora | Ecdysozoa | Velvet worms |
| Tardigrada | Ecdysozoa | Water bears |
| Nematoda | Ecdysozoa | Roundworms |
| Nematomorpha | Ecdysozoa | Horsehair worms |
| Kinorhyncha | Ecdysozoa | Mud dragons |
| Priapulida | Ecdysozoa | Penis worms |
| Loricifera | Ecdysozoa | |
| Echinodermata | Deuterostomia | Starfish, sea urchins, sea cucumbers |
| Hemichordata | Deuterostomia | Acorn worms |
| Chordata | Deuterostomia | Sea squirts, amphioxus, fish, humans |
| Chaetognatha | Lophotrochozoa/ Ecdysozoa | Arrow worms |
| Acoelomorpha | Uncertain | |
| Xenoturbellida | Uncertain | |
| Orthonectida | Uncertain | |

# Chapter 3
# **The evolutionary tree of animals**

> The time will come, I believe, though I shall not live to see it, when we shall have very fairly true genealogical trees of each great kingdom of nature.
>
> Charles Darwin, letter to T. H. Huxley (1857)

## Building a tree of life

Darwin realized that a branching tree was a good metaphor for describing the course of evolution. In 1837, he made a small sketch of an evolutionary tree in one of his personal notebooks, with the tantalizing words '*I think*' written above. The concept would have come quickly to Darwin once he realized that one species could give rise to two or more 'daughter' species – a process known as speciation. Evolutionary trees, or phylogenetic trees as they are also called, are simply diagrams that depict these speciation events. Every branching point on a phylogenetic tree, where one line forks to give two lines, is a visual portrayal of one species becoming two.

Phylogenetic trees are easy to comprehend when they include similar animal species. For example, if one line on a tree leads to the Large White butterfly (*Pieris brassicae*) and another leads to the Small White butterfly (*Pieris rapae*), the point where these two lines meet marks the speciation event separating these two very similar butterflies. This is the point in history when two populations of their 'common ancestor' became separated such that they could no longer interbreed. Importantly, these two populations will not yet have acquired the distinct characters of the two species; indeed, they may look essentially identical. But very often phylogenetic trees do not just contain closely related species; they may depict the evolutionary relationships between large sets of animals, such as between insects, spiders, snails, jellyfish, and humans. These trees should be viewed in exactly the same way. If one line on the tree leads to insects and another leads to spiders, where these two lineages meet marks the long-extinct common ancestor of the two groups. That ancestor was neither insect nor spider, and when it underwent speciation, it gave rise to the almost indistinguishable ancestors of these two groups.

Although Darwin sketched the idea of a tree in his notebook, and enlarged upon it in his only illustration in *The Origin of Species*, he did not try to resolve who was actually related to whom. For Darwin, evolutionary trees were just a concept: a way of thinking about evolution. Many subsequent evolutionary biologists did attempt to put names onto branches of the tree. It is an important problem and one that should be soluble. After all, there should be one single tree of animal life, depicting the true course of animal evolution. Any drawing of a phylogenetic tree is therefore a clear and explicit hypothesis about the pathway followed in evolution. Some of the earliest evolutionary trees were drawn by the German zoologist Ernst Haeckel in the 1860s and 1870s. Many of Haeckel's trees were striking in their artistic detail, complete with knotted bark and twisted branches, and the names of particular animal groups at the end of each twig or leaf. Haeckel based his trees, and therefore his hypotheses about animal evolution, on several lines of evidence, but he especially liked to use characters from embryology. This was partly because he thought that embryos change slowly in evolution. Also, even when adults look very different, similar features can sometimes be found in their development. Some of Haeckel's conclusions are still compatible with modern ideas, such as his placement of jellyfish and sea anemones on a branch that split early from the rest of animal life. Other ideas seem surprising to us today and are certainly incorrect, such as placing echinoderms (starfish and sea urchins) as a branch close to arthropods such as insects and spiders.

Over the next 80 years, zoologists made better descriptions of the anatomy of animals and studied their embryonic development in more detail, with attention being paid to the great diversity of invertebrate phyla. But even by the middle of the 20th century, no clear consensus had been reached. There was no single agreed phylogeny of the Animal Kingdom. Each author would give a slightly different evolutionary tree, although certain relationships were always present. One scenario, outlined below, became prevalent particularly in American textbooks and has been called the 'Coelomata hypothesis'.



## The Coelomata hypothesis

In this evolutionary tree, the main lines of evidence used to decide which animal phyla are most closely related were symmetry, germ layers, body cavities, segmentation, and patterns of cell division in the early embryo. Most familiar animals, including worms, snails, insects, and ourselves, have only one mirror-image plane, or axis of symmetry. This runs in the head-to-tail direction, separating the left-hand side from its mirror image on the right-hand side of the body. There are many deviations from precise symmetry, such as coiled shells in snails, lopsided claws in crabs, or placement of the human heart on the left-hand side of the body, but these are minor modifications. Fundamentally, most animals have near mirror-image left- and right-hand sides – an organization called 'bilateral' symmetry. In contrast, four animal phyla have no clear head and tail ends, and no left or right sides. These 'non-bilaterian' phyla, or basal animal phyla, have either no symmetry or radial symmetry. They comprise the Cnidaria (jellyfish, sea anemones, corals) and Porifera (sponges), plus two less well-known groups called Ctenophora (comb jellies) and Placozoa.

The second line of evidence was the number of 'germ layers'. Germ layers are layers of cells that arise early in the embryo and become more complex during development. Most animals have three germ layers, with the inner layer (endoderm) forming the wall of the gut, the outer layer (ectoderm) forming skin and nerves, and the middle layer (mesoderm) developing into muscle, blood, and other tissues. However, the non-bilaterian or basal phyla have only two germ layers (ectoderm and endoderm), at least to a first approximation. Whether there is something similar to mesoderm in these animals is controversial. Because of these two lines of evidence, symmetry and germ layers, the bilateral animals were placed in one large group, called the 'Bilateria' or 'bilaterians' (also called 'triploblasts' on account of the three germ layers), with the other phyla arising from branches that separated earlier in animal evolution.



Moving onto the bilaterians, one character given special attention in the Coelomata phylogeny was the presence or absence of fluid-filled spaces within the body. The embryos of some bilaterians, most notably annelids (such as earthworms) and molluscs (for example, slugs and snails), have large fluid-filled body cavities, lined with leak-proof epithelial cell sheets. The embryos of chordates, such as humans, also have these cavities, as do embryos of echinoderms (starfish and sea urchins). A body cavity of this type is called a coelom; these animal phyla were therefore called 'coelomates', and they were grouped near each other in the evolutionary tree. In earthworms, the coeloms persist into the adult, where they act as a liquid skeleton. In some other animals, including arthropods (such as insects and spiders), the coeloms can be very small or may disappear later in development, but these animals were still placed within the coelomate part of the tree. (In some other phylogenetic trees, only some of the coelomates were grouped together.) Another reason why arthropods were placed close to annelids is that both types of animal have bodies divided into repeating articulated units or 'segments'. The segmentation can be seen very clearly in the body of a centipede or an earthworm as a series of rings around (and inside) the body. In many trees, therefore, a particular super-branch of 'segmented coelomates' was defined, and called the Articulata.

In contrast to coelomates, there are also bilaterians in which the mesoderm remains solid, without any fluid-filled cavity. These animals were called acoelomates and include Platyhelminthes (flatworms, flukes, tapeworms) and Nemertea (ribbon worms). Intermediate between these two categories were the pseudocoelomates, such as Nematoda (roundworms), which have poorly defined body cavities without an epithelial cell layer. An assumption of the Coelomata phylogeny was that all the coelomates grouped together, and that the acoelomates split off earlier. The acoelomates were thought to be ancestors of the coelomates, and thus the most 'primitive' of the bilaterian animals. Another implication was that there had been a rise in complexity through bilaterian evolution, from acoelomate to coelomate, possibly via pseudocoelomate, neatly mirrored by the animal phyla surviving today.



## A new tree of animals

Not every zoologist adhered to the above view, but it remained a popular hypothesis for several decades. The most prevalent alternative theory split the Bilateria into two main groups (Protostomia and Deuterostomia) and gave less attention to body cavities; however, it still used segmentation to group arthropods and annelids into Articulata. In 1988, however, a new line of evidence was brought to bear on the question, and this quickly gave an indication that something might be very wrong with the Coelomata hypothesis, and also with the idea of Articulata. A team of researchers at Indiana University in the USA, headed by Rudolf Raff, set out to use gene sequence data to investigate the evolutionary relationships between animal phyla. Genes accumulate mutations over time, so differences in DNA sequence between species should reflect how long it has been since they shared an ancestor. Animal phyla that are close relatives would have similar DNA sequences for a particular gene; more distant relatives would have more different DNA sequences. Raff and colleagues focused attention on the gene coding for small subunit ribosomal RNA, one of the components of the ribosome, a structure found in all cells. The main advantage of this gene is that it is present in every animal species, doing the same job: helping make proteins.

The 1988 study marked the beginning of a revolution in using DNA sequence information to hunt for the true phylogenetic tree of animals. Even though the technology was new and the analysis methods in their infancy, one conclusion seemed clear right from the start. The segmented annelids and segmented arthropods had very different ribosomal RNA gene sequences; there was no evidence supporting the Articulata group. Over the next 20 years, the DNA sequences of many more genes were determined, from many more species, and computer-based analysis methods were refined and improved. The most reliable phylogenetic trees now include over a hundred genes from each animal, and they show a remarkably consistent picture. This 'new animal phylogeny' has some similarities with the older trees, but it also has some key differences.



In the new animal phylogeny, the four non-bilaterian phyla branch off the main tree early, just as they did in the Coelomata and other morphology-based phylogenies. This implies that germ layers and symmetry were giving an accurate picture. Jellyfish, sea anemones, corals, comb jellies, and sponges are indeed 'basal' animals. After these basal phyla diverge, the rest of the animals comprise the Bilateria. It is within this bilaterian group where the hypotheses differ. For example, in the new animal phylogeny there is no region of the tree composed only of acoelomates, no section for pseudocoelomates, and no grouping of just coelomates. Instead, all three types of body organization are mixed up.

Since coeloms are found in several different parts of the 'new' tree, this implies either that body cavities arose more than once in evolution, or that they can be lost, or both. From a functional point of view, this is perhaps not surprising. Fluid-filled cavities provide an advantage to invertebrates living in many environments – they provide support to the body and act as an incompressible bag against which different sets of muscles can squeeze. For soft-bodied animals, this increases the power and efficiency of animal movement, allowing burrowing, faster crawling, and even occasionally swimming. From the perspective of drawing evolutionary trees, it means that body cavities were poor markers of relatedness. The same is true for segmentation. Dividing the body into units offers advantages in certain environments, for example by increasing the efficiency of movement, and it also probably evolved more than once. Segments, like coeloms, come and go too readily in evolution to be markers of who is related to whom. It seems there is no grouping called the Coelomata, and no Articulata.

So what shape is the evolutionary tree built from DNA sequences? In the new animal phylogeny, rapidly gaining wide acceptance, the Bilateria divide into three great groups, which might be called 'superphyla'. Each of these contains several phyla. The superphylum to which we belong is called the Deuterostomia. Along with our phylum the Chordata, this deuterostome group includes the Echinodermata (starfish and sea urchins) and the Hemichordata (including the foul-smelling acorn worms). The older phylogenies almost always had a group called Deuterostomia as well, but it usually also included a few other animals which have now been moved elsewhere on the basis of the DNA data, notably the Chaetognatha, or arrow worms.

The other two great superphyla of bilaterians were a surprise. They had not been suspected from comparison of anatomy, and were not present in any of the older, traditional trees. Nonetheless, each is now strongly supported by DNA sequence data. Since they were proposed only recently, these two groups of animals needed new names. They are each a bit of a mouthful. One, containing Arthropoda (insects, spiders, crabs, centipedes), Nematoda (roundworms), and several other phyla, is called 'Ecdysozoa'. The other, containing Annelida (earthworms, leeches), Mollusca (snails, octopus), Platyhelminthes (flatworms, flukes, tapeworms), Bryozoa (moss animals), and others, delights in the name 'Lophotrochozoa'.

A phylogenetic tree is best shown using a diagram. As summarized in Figure 2, the new animal phylogeny has the four non-bilaterian



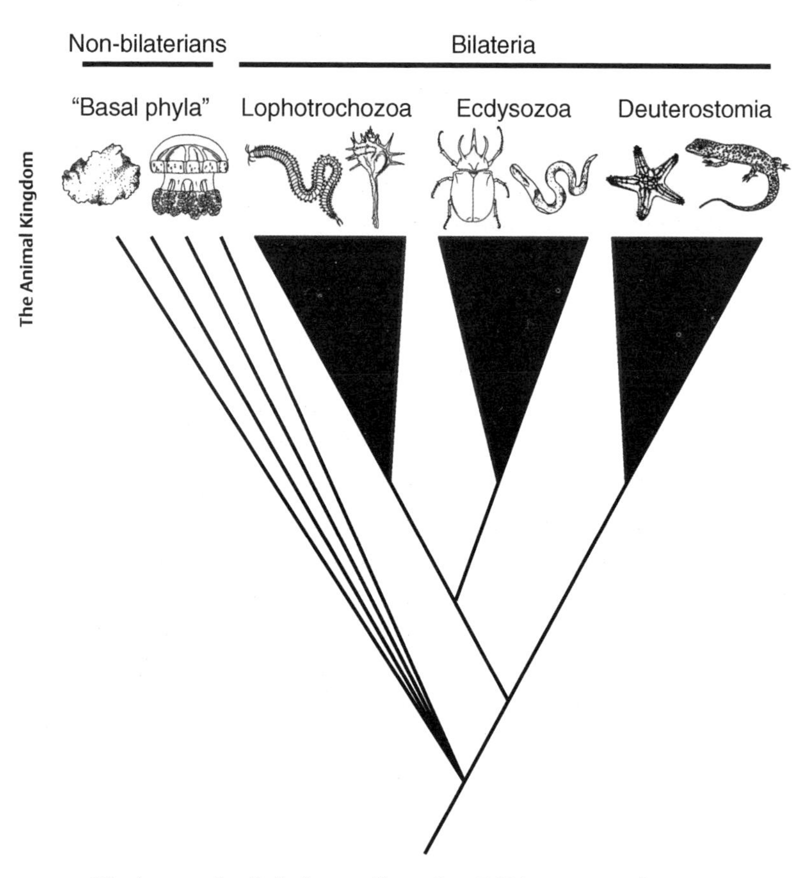


**2. The 'new animal phylogeny' based on DNA sequence data**

phyla branching early in animal evolution, leaving the large group of Bilateria. The bilaterians then divide into the three great superphyla, Deuterostomia, Ecdysozoa, and Lophotrochozoa, as described. Incidentally, the latter two groups are closest to each other, and approximate to the 'Protostomia' of some older trees. It is important to realize that among the three great groups, none is 'higher' or 'lower' than any other, since all still exist today. There is no rising Scale of Nature. In the remaining chapters of the book, we will look at the animals in each of these branches, starting with the non-bilaterian phyla, and then dealing with the three large bilaterian superphyla in turn. The order is arbitrary. Just because humans sit within the Deuterostomia does not give our group any special priority in the tree.

# Chapter 4
# Basal animals: sponges, corals, and jellyfish

> The bottom was absolutely hidden by a continuous series of corals, sponges, actiniae and other marine productions, of magnificent dimensions, varied forms and brilliant colours.... It was a sight to gaze at for hours, and no description can do justice to its surpassing beauty and interest.
>
> Alfred Russel Wallace, *The Malay Archipelago* (1869)

## Porifera: the sponges

Sponges are the least animal-like of all members of the Animal Kingdom. Most sponges are vase-shaped, but some appear as lumpy, irregular growths encrusting rocks in the sea, or pebbles and fallen branches in lakes and rivers. For these animals, the concepts of front and back, top and bottom, or left and right, do not apply rigidly. They have no clear nerve cells or muscles, but they can move, very slowly, and – like other animals – sponges can respond to touch and can sense chemical changes in their environment. Unlike other animals, they do not have a true mouth or gut, but instead use a complicated system of water flow to capture food. Sponges can be recognized by the presence of one or more large pores or holes on their surface, although there are also thousands of much smaller pores. A constant stream of water flows into the small holes and out of the larger ones. This water current, which carries dissolved oxygen and particles of food such as bacteria, is set up by an important type of cell found lining a network of hollow canals and cavities within the sponge. These feeding cells, or choanocytes, have a beating flagellum and resemble the unicellular collared flagellates encountered earlier. They function in a different way, however, since – unlike collared flagellates – the choanocytes do not catch food using their collar as a simple net. Instead, the chambers containing choanocytes have a larger cross-sectional area than the pores, meaning that the water flow slows greatly once it is drawn into the sponge. With the incoming water flow now almost stationary, sponge cells can engulf bacteria and other particles of food.

Although sponges have many different types of cells, most are not organized into organs with discrete functions, like kidneys, livers, or ovaries (although the choanocyte chambers could be considered simple organs). For this reason, sponges are sometimes described as having a 'tissue-level' organization. Some sponges have astonishing powers of regeneration, so extreme that they were the inspiration for regenerating aliens in the science-fiction television series *Doctor Who*. The defining experiments that revealed this property were published in 1907 by Henry Van Peters Wilson of the University of North Carolina, USA. Wilson mashed up a living sponge and passed it through a fine cloth, the sort used for sieving flour, thereby splitting most of it into individual cells. Wilson then observed that these cells gradually crawled back together and reassembled into a new sponge! Furthermore, if the cells of two different species were mixed together, they would sort themselves out and regenerate into the two original sponges again. Although regeneration is found in many branches of the Animal Kingdom, no other animals are as expert as some of the sponges.

In between the outer and inner layers of cells, sponges have 'connective tissue' reinforced with tough fibres of a protein called spongin or with minute spears or stars (spicules) made of calcium carbonate or silica. The former type of sponge, with a spongin skeleton but no spicules, was the source of the old-fashioned bath sponges widely used for washing and cleaning, although now largely replaced by synthetic foams. Examples include the genera *Spongia* and *Hippospongia*. The collection and use of sponges dates back many centuries. In the 1st century AD, Pliny the Elder described in detail how to use sponges to clean wounds, reduce swellings, arrest bleeding, and treat stings. Even earlier, in the 4th century BC, Aristotle described which type of sponge should be used to line helmets, writing:

> the sponge of Achilles is exceptionally fine and close-textured and strong. This sponge is used as a lining to helmets and greaves, for the purpose of deadening the sound of the blow.

Remarkably, it is not only humans who use sponges as tools. In Shark Bay on the Western Coast of Australia, a population of bottlenose dolphins has learned to snap off pieces of living sponge and attach these to their snouts to protect themselves when foraging for food in the sandy bottom.

The sponges comprise a phylum, the Porifera, which is in turn divided into three classes: Demospongia (including bath sponges), Calcarea (with calcium carbonate spicules), and the rare, deep-sea-dwelling Hexactinellida. The hexactinellids, also known as glass sponges, are particularly beautiful and have some important differences from the other sponges. One peculiar feature is that much of their body is made of 'syncytia': sheets of cytoplasm containing many nuclei, not separated into individual cells by membranes. Also unusual is that their silica spicules are woven together into delicate lattice-like structures, like elaborate cages of glass. The best-known example is 'Venus's flower basket', *Euplectella aspergillum*, which lives attached to rocks on the Pacific Ocean floor, and has a cylindrical 30-centimetre tower-like skeleton made of intricately laced glass fibres. Often, it is found with a pair of live shrimps, male and female, trapped within the glass fibre cage, having grown too large to swim out through the gaps in the sponge's skeleton. The shrimps' offspring, however, can escape through the latticework walls and swim off to populate other Venus's flower baskets, leaving the two parent shrimps behind in their permanent partnership. In an old Japanese custom, specimens of this sponge were often given as wedding gifts symbolizing eternal union.

## The strange case of *Trichoplax*

Sponges are not the only animals to lack the three defined axes: head to tail, top to bottom, and left to right (bilateral). Three other phyla are also 'non-bilaterian' in their organization: the Cnidaria (sea anemones, corals, and jellyfish), the Ctenophora (comb jellies), and the Placozoa. Originally, only a single species was placed in the last of these phyla, a tiny pancake-shaped creature called *Trichoplax adhaerens*, meaning 'sticky, hairy plate'. Recent genetic analysis suggests it is not alone, however, and there are actually several similar species of these minute creatures, crawling and floating in tropical and subtropical seas from the Pacific Ocean to the Caribbean, and the Mediterranean to the Red Sea. At first sight, *Trichoplax* could easily be confused for a very large amoeba, between half and one millimetre across, but on closer examination it can be seen to be made of thousands of cells, as befits a true animal. Its irregular flattened shape has no preferred front end, and it crawls along hard surfaces in any direction through a combination of shape changes and the beating of thousands of microscopic cilia covering its underside. With no mouth or gut, *Trichoplax* feeds by secreting enzymes from its undersurface, which break down food matter, such as single-celled algae, into nutrients for absorption. All in all, placozoans are very unusual animals and they have long puzzled zoologists.

They were first discovered in 1883 by the German zoologist and sponge expert Franz Eilhard Schulze but, interestingly, he did not actually discover *Trichoplax* in nature. Schulze found the new animal crawling on the glass walls of a marine aquarium in Austria, meaning that at first there was no clue as to where it lived

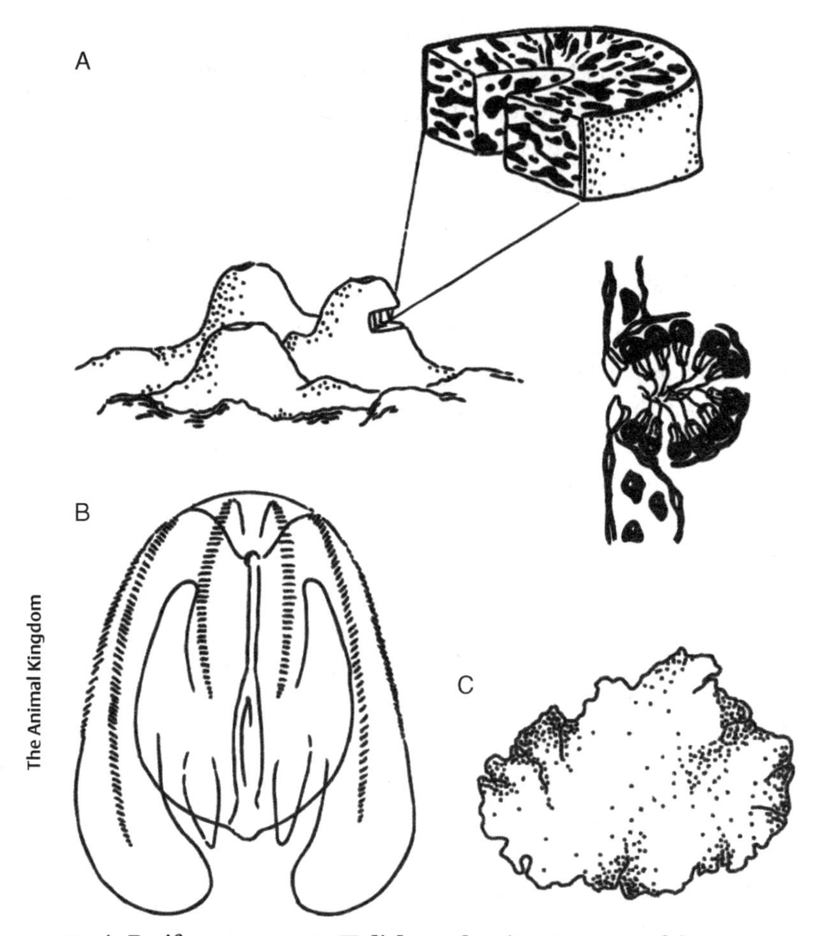


**3. A, Porifera, or sponge: *Haliclona*, showing structure of choanocyte chamber; B, Ctenophora, or comb jelly: *Mnemiopsis*; C, Placozoa: *Trichoplax***

in the wild. Indeed, many zoologists later claimed that Schulze was mistaken in describing *Trichoplax* as a new animal at all, arguing that it was simply the larva of a well-known sea anemone-like animal. It was almost a century before Schulze was fully vindicated, as extensive research on placozoans in the wild and in the laboratory have now proved them to comprise a distinct phylum in their own right, albeit one rather short on species numbers.

Those wishing to follow Schulze should be warned, however. I was once forcibly told to leave an aquarium shop when the irate manager found me peering through a magnifying glass at the scum in his fish tanks.

## Ctenophora: comb jellies

The Ctenophora comprise a third phylum of non-bilaterian animals, and one very different in body organization from either sponges or placozoans. Also known as comb jellies, the ctenophores are slow-moving predators that drift through the seas eating other slow-moving animals, such as other ctenophores, crustaceans, and marine larvae. Unlike most predators, comb jellies do not chase or stalk their prey. They simply bump into small planktonic animals and ensnare them with tiny drops of glue secreted by specialized cells, usually concentrated along two long tentacles trailing away from the sides of the mouth. Ctenophores, in contrast to sponges and placozoans, have nerve cells and a balance sense organ so they can interact with their environment rapidly and responsively. Although most are essentially blobs of jelly just a few centimetres in size, almost everyone who has seen one alive would rank comb jellies amongst the most beautiful animals on the planet. Their most conspicuous features are the eight 'combs' that run as strips along the body, each containing thousands of cilia. The cilia beat in a highly coordinated way, with each one beating just after its neighbour to form a set of 'metachronal waves', rather like the 'Mexican waves' that occasionally circulate around a football stadium during lulls in play. This gentle wafting of thousands of tiny cilia propels the animal slowly and silently through the sea, but also scatters light to create a shimmering rainbow of colours, constantly changing and flickering. The best-known comb jellies are the grape-sized 'sea gooseberries' such as *Pleurobrachia* found throughout Pacific and Atlantic oceans, and around the British coast. But the most spectacular comb jelly is undoubtedly the giant 1-metre-long *Cestum veneris*, or Venus's girdle, named after the Roman goddess of love. Instead of the usual egg-shape typical of ctenophores, this striking, iridescent animal has an elongated, ribbon-like body that shimmers in the sea as the sun's rays are scattered by its rows of cilia. In Richard Dawkins's words, *Cestum* is 'too good for a goddess'.

Most comb jellies have little direct effect on humans, except through a minor role in marine food webs. One species, however, stands apart as the villain of the basal invertebrates. In the 1980s, the Atlantic comb jelly *Mnemiopsis* was accidentally introduced into the Black Sea, probably in ballast water carried by commercial shipping. Once in its new environment, away from natural competitors and predators, it reproduced rapidly – all the while consuming vast quantities of larval fish and crustaceans. Some (controversial) estimates placed the total seething mass of diminutive comb jellies in the Black Sea at over half a billion tonnes. The local anchovy fishery, already under heavy fishing pressure, was decimated. While ecologists debated what to do, a possible solution arrived, unplanned, in the shape of another accidental introduction. The newcomer was a second comb jelly, this time the voracious *Beroe*. Fortuitously, *Beroe* does not eat fish or crustaceans, but is instead a specialist predator of other comb jellies and nothing else. As the invading *Beroe* now feast on the *Mnemiopsis*, the fish stocks are showing gradual and welcome signs of recovery.



## Cnidaria: stings and super-organisms

Of the four 'non-bilaterian' phyla, sponges and placozoans lack precise symmetry, while comb jellies have 'biradial' symmetry, meaning that their bodies are symmetrical by a 180-degree rotation. The fourth and largest of the non-bilaterian phyla, the Cnidaria, contains some very familiar animals including jellyfish, sea anemones, and corals. The bodies of these animals also lack head-to-tail, top-to-bottom, and left-to-right axes, and with few exceptions they also have radial or rotational symmetry. The basic structure of a cnidarian body is cup-shaped or flask-shaped, with a single large opening at one end that acts as both mouth and anus. Tentacles surround this opening, each armed with thousands of stinging cells called cnidocytes. These cells, which fire tiny barbed harpoons or nematocysts laced with poison within three milliseconds of being touched, are the cnidarian's chief weapon of attack and defence.

Cnidarians have nerve cells, and as in ctenophores, these are arranged in a lattice-like network around the body rather than being organized into a single distinct brain and 'central nerve cord' as in most other animals. The cell layers that form the body, the ectoderm on the outside and endoderm on the inside, are separated by a substance called mesoglea. Although much of the mesoglea is made of proteins rather than layers of living cells, in many cnidarians it has scattered cells crawling within it, and in some species it even has muscle cells organized into contractile fibres. However, the cells within the mesoglea do not form complex organs, and so cnidarians are usually described as having a body built from just two basic cell layers.

The cnidarians divide into four groups. The first, the anthozoans, includes sea anemones, such as the brightly coloured Snakelocks and Beadlet anemones found in rock pools. In these animals, the single opening to the body faces upwards, while the opposite end sticks rather weakly to the rocks. Once the incoming tide has covered them, sea anemones open their crown of tentacles and wait for small prey animals to drift or swim near them, whereupon these will be promptly stung and eaten. Although generally static animals, sea anemones are not totally fixed but can detach and move to a new location by drifting or gentle swimming. They can also creep slowly along on their single adhesive foot, sometimes to find a more favourable location for feeding and sometimes to engage in ferocious slow-motion battles when two sea anemones attempt to sting each other with inflated clubs armed with nematocysts.

Corals are also anthozoans and they demonstrate a character that has arisen repeatedly in animal evolution: coloniality. Coral consists of thousands or even millions of tiny animals, each like a miniature sea anemone just a few millimetres across, but interconnected to make a giant 'super-organism'. A living coral grows by budding of the tiny 'zooids' such that the whole colony has the same genetic make-up. It is one large clone. In some species, the colony resembles a fan; in others, it branches like deer's antlers; and yet others grow to look like grisly body parts such as brains or 'dead-men's fingers'. The most impressive of all, however, are the reef-building corals which secrete calcium carbonate around the budding zooids to form giant chalky structures upon which many other animal species can also make their home.



The second group of cnidarians, and looking superficially like sea anemones, are the hydrozoans. These include some large and colourful marine species, plus the diminutive hydra found in ponds and rivers. Named after a multi-headed water beast of Greek mythology, the hydra body is a miniature tube a few millimetres in length, with an upward-facing mouth surrounded by stinging tentacles. All species of hydra will catch and eat tiny fresh-water invertebrates, but several species back this up with an additional trick. The 'green hydra' has enslaved a unicellular alga, which grows inside the hydra's gut cells, giving the whole body a bright green appearance and providing the hydra with food through photosynthesis. Just as some anthozoans live in interconnected colonies, so too with some hydrozoans. The infamous Portuguese Man o'War *Physalia physalis* is a giant colonial hydrozoan comprising a gas-filled float beneath which are thousands of connected zooids dangling menacingly in 10-metre strings, bristling with venomous nematocysts.

When a cnidarian has a 'mouth up' orientation, as in adult sea anemones, corals, and hydra, it is called a 'polyp'. The opposite body orientation, with the opening facing downwards, is known as a 'medusa' and is the form typically seen in scyphozoans or jellyfish. Many cnidarians have life cycles that alternate between the two orientations, mouth up and mouth down. There are additional differences beyond simply orientation, and recent research using gene expression patterns has shown that downward-pointing medusa tentacles are not actually the same structures as upward-pointing polyp tentacles. Jellyfish, like all cnidarians, are predators. These bell-shaped gelatinous animals drift or gently swim through the upper waters of the sea, propelled by rhythmic pulsations of their body wall. Surface waters of the oceans teem with planktonic life, such as crustaceans and immature fish, which the jellyfish ensnare with trailing tentacles armed with poisonous nematocysts. Many swimmers have had first-hand experience of accidentally brushing against jellyfish tentacles and suffering a painful rash. Several variations on the basic jellyfish theme have evolved, one of the most unusual being seen in animals of the order Rhizostomae. In these jellyfish, there is no single downward-pointing mouth, since this is closed off by fused tissue, and instead there are a multitude of tiny mouth-like openings on eight branching arms, each connected to the gut by a complicated canal system. Many rhizostomids, such as *Mastigias papua*, supplement their food intake by harbouring in their tissues millions of symbiotic algae capable of producing energy by photosynthesis. This enables *Mastigias* to live at incredibly high densities. In 'Jellyfish Lake' on the Pacific island of Eil Malk in Palau, intense aggregations of *Mastigias papua* can sometimes reach a thousand 6-centimetre animals per cubic metre of sea water.

More dangerous to humans than true jellyfish, or even the Portuguese Man o'War, are the animals comprising the fourth group of cnidarians: the cubozoans. Also known as box jellies on account of their shape, these animals are commonest in coastal tropical seas. Unlike true jellyfish, each cubozoan has 24 eyes, including 6 with lens, iris, and retina capable of forming an image of distant objects. Some species, such as the Australian sea wasp

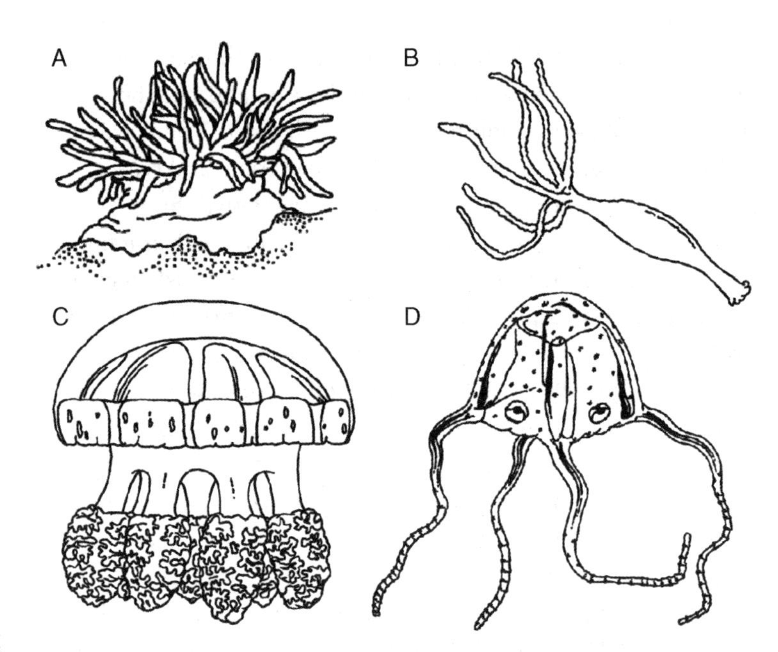


**4. Phylum Cnidaria: A, Anthozoan, sea anemone; B, Hydrozoan, *Hydra*; C, Scyphozoan, or jellyfish, *Mastigias papua*; D, Cubozoan, or box jelly, *Carukia barnesi***

*Chironex fleckeri*, are deservedly feared by swimmers on account of their extremely powerful venom. Sea wasp stings are intensely toxic, and can be fatal even for humans. The stings from some other cubozoan species are less painful on contact, yet can trigger an unusual delayed reaction known as 'Irukandji syndrome', named after the Australian Aborigines from the north Queensland coast where box jellies are common. Swimmers stung by Irukandji cubozoans gradually develop excruciating back pain, muscle cramps, nausea, increased blood pressure, and a range of disturbing psychological effects including 'a feeling of impending doom'.

# Chapter 5
# The bilaterians: building a body

Man is but a worm.

Edward Linley Sambourne, *Punch Magazine* (1881)

## Life with a front end

You are a bilaterian. So are fish, birds, worms, squid, cockroaches, and millions of other animals. In fact, most animals are Bilateria. As the name indicates, this vast division of the Animal Kingdom comprises the animal phyla with 'bilateral symmetry', which means that these animals have just a single line of mirror-image symmetry running right down the centre of the body. This symmetry line separates the left- and right-hand sides of the body, and by implication there must then be distinct front and back ends, and top and bottom surfaces, which are not symmetrical. In humans, the left- and right-hand sides are just as you recognize them, but the front end (anterior) of your body is actually your head, the back end (posterior) of your body is what you sit on, your top or 'dorsal' surface is found along your spine, while your bottom or 'ventral' surface is your belly. These orientations make sense when we remember that humans stood upright only recently, in evolutionary terms.

Bilateral symmetry contrasts with the rotational symmetry found in most cnidarians and comb jellies, and the lack of clear symmetry seen in placozoans or sponges. The similarity in body organization amongst the bilaterians is more than skin deep. The bilaterian animals have well-defined blocks of muscle which can be used for active movement, and almost all have centralized nerve cords with an anterior brain, plus specialized sense organs concentrated at the front end. Most have a tube-like or 'through-gut' with a separate mouth and anus, allowing efficient processing of food, and the exceptions, with just a single opening to the gut, could have reverted to this condition secondarily. The evolution of bilaterian animals marked the rise of animals with active, powerful, and directed locomotion, able to burrow, crawl, or swim while facing their environment head-on with batteries of sense organs, leaving their waste products behind them. The bilaterians truly explore and exploit the world in three dimensions.



The distinctions between bilaterians, or triploblasts as they are also known, and the more 'basal' animal phyla have been noted for over a century. In 1877, the influential English zoologist Ray Lankester contrasted the embryos of bilaterians with those of cnidarians and sponges, and pointed out that in their early development bilaterians have an extra layer of cells destined to develop into the well-defined muscle blocks of the adult. The similarities seen within embryos and in body symmetry are certainly fundamental. Towards the end of the 20th century, biologists were astounded to find similarities that went far, far deeper – right down to the DNA. The discovery that all bilaterians use the same set of genes to build their bodies represents one of the most fascinating scientific breakthroughs of the 20th century, and one that has changed biological science from the 1980s onwards. It was a discovery with explosive impact, yet this was a revolution with a slow fuse.

## Homeosis and Hox genes

William Bateson is remembered today as one of the founders of the science of genetics. Long before he became famous, and soon after completing his undergraduate degree, the young Bateson published a series of scientific papers on the anatomy of the acorn worm, a marine invertebrate whose evolutionary position was then unresolved. Bateson's work was received with some acclaim, yet he was not satisfied, saying that it gave little insight into how evolution actually worked. In a letter to his mother, Bateson wrote:

> Five years hence no-one will think anything of that work, which will be very properly despised. It hasn't any bearing whatever on the things we want to know. It came to me at a lucky moment and was sold at the top of the market.

What Bateson really wanted to know was how variation arose within species. So, for the next eight years, Bateson devoted himself to cataloguing 'variants' in animals and plants, publishing in 1894 his magnum opus *Materials for the Study of Variation*. Amongst the many gems in this book, Bateson discussed an unusual type of variation in which an animal is found with one 'structure' replaced by another that would usually be found elsewhere in the body, such as an antenna growing where an eye should be. These strange 'homeotic' variants remained little more than curiosities until 1915 when Calvin Bridges showed that one such change in a fruitfly was passed on to the fly's offspring. The inheritance was key. It pointed the finger at genes. The implication was that there must be genes instructing body parts to develop correctly, and when one of these genes has an error – a mutation – the instructions are misread. One region of the body will develop as if it is a different region. In this first homeotic mutation, wings, or rather parts of wings, grew where they should not be. Bridges called the mutation *bithorax*. Oddities such as *bithorax* are far too dramatic to play any direct part in evolution; flies with such big changes to anatomy would not survive in nature. But the mutation gives a clue to how genes build bodies, and that certainly is relevant to understanding how animal evolution works.

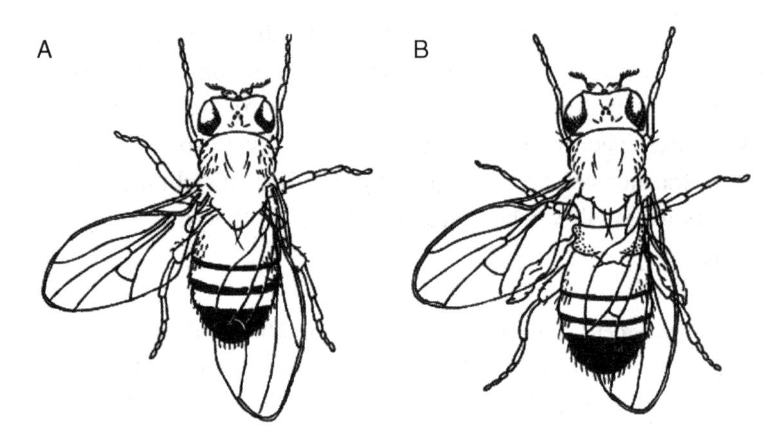


**5. A, Normal fruitfly; B, Calvin Bridges's *bithorax* mutation**



The finding was taken up and pursued with great energy and persistence by another geneticist, Ed Lewis. In a series of brilliant papers, including a Nobel Prize-winning masterpiece in 1978, Lewis showed that *bithorax* was not alone. There were several genes that could mutate to cause homeotic mutations, with each mutation affecting a different region of the fly's body, and all mapping to one part of one fruitfly chromosome. Another geneticist, Thom Kaufmann, found genes that controlled development of the head and front of the body, and so a picture emerged of a whole set of 'homeotic genes', each telling cells in the embryo where they were. The homeotic genes act like postcodes, telling cells where they are along the head-to-tail axis of the fly.

When their DNA was analysed, all homeotic genes turned out to be similar to each other, particularly along a stretch of 180 base pairs. This region, which became known as the 'homeobox', was a molecular signature of homeotic genes – or Hox genes as they were soon named. Homeobox sequences were also found in the DNA of some other fruitfly genes, but always in genes that had something to do with controlling development, such as the gene *fushi tarazu* involved in formation of the fly's segments. But very soon, this became a story with relevance far beyond fruitflies. Few biologists were prepared for the impact of what was unravelling. The year was 1984, the centre of activity was Basel, Switzerland, and a dynamic team of researchers including Bill McGinnis, Mike Levine, Atsushi Kuroiwa, Ernst Hafen, Rick Garber, Eddy De Robertis, Andres Carrasco, and Walter Gehring were pushing back the boundaries of biological knowledge. McGinnis and colleagues tested whether homeobox sequences could be detected in DNA extracted from other animals, with striking results. Not only did other insects have homeoboxes, but the first experiments suggested that perhaps so did worms, snails, and possibly even mice and humans! Carrasco, McGinnis, Gehring, and De Robertis quickly isolated and sequenced the DNA for a frog homeobox gene and proved the point: vertebrates really do have homeobox genes.

The buzz around the scientific community was electrifying. The top journals clamoured to showcase each new finding, and each published paper was devoured by an eager readership. Every conference and seminar was dominated by homeoboxes. After one scientific lecture in London at which a colleague reported a new developmental control gene, I recall the first question asked was 'Does it have a... shall I say the magic word?' I even knew several scientists who just stopped their life's research there and then, and started afresh to work on homeobox genes.

The discovery of the homeobox, and the finding that homeobox genes are present in animals as diverse as flies and frogs, started a revolution. Prior to 1984, there was essentially no knowledge of how genes controlled body patterning in most species. Perhaps homeobox genes provided a new way into the problem? This question prompted Jonathan Slack to liken the discovery of the homeobox to the finding of the ancient Rosetta Stone, unearthed in Egypt in 1799, which provided the first translation between ancient scripts. In the same way, did we now have a way of comparing the control of embryonic development between widely different species? The optimism was not shared by everyone, but it turned out to be well founded. Many of the homeobox genes in vertebrates, such as frogs and humans, are indeed equivalent to the fruitfly's homeotic or Hox genes, and they play essentially the same roles. As in flies, our own Hox genes act as postcodes, telling human cells where they are along the head-to-tail axis.

For evolutionary biology, the implications were immense. If vertebrates and insects have Hox genes, then surely by implication so must all Bilateria? If vertebrates and insects use these genes to denote position along the main axis, then this property too must date back to the origin of the bilaterally symmetrical animals. We can be confident in these statements simply because the common ancestor to flies and humans was also the common ancestor to all of the Ecdysozoa, Lophotrochozoa, and Deuterostomia. One bilaterian phylum might possibly have branched off a little earlier, the Acoelomorpha, but recent evidence suggests that even in these animals the Hox genes play a similar role. So this entire swathe of the Animal Kingdom, the 29 animal phyla with a clearly distinct front end and back end, the active explorers of the three-dimensional world, use the same set of genes to pattern the main head-to-tail axis.



## Up and down, left and right

What of the other two body axes: top to bottom and left to right? Here too, genes have been found that ensure cells know where they are. Furthermore, just as was shown for Hox genes, it turns out that very different bilaterian animals use essentially the same genes – but with an interesting twist. In the embryo of a fly, cells on the bottom or ventral side will form the main nerve cord, with the gene *sog* playing a key role. Cells on the top or dorsal side form epidermis, with this opposite fate controlled by the gene *dpp*. Vertebrates also have *sog* and *dpp* genes, although known by different names. The vertebrate *sog* gene, called *chordin*, is expressed on the side destined to become dorsal, where our nerve cord lies; *BMP4*, one of the vertebrate *dpp* genes, marks ventral. In terms of orientation, it is simply the opposite way round to flies.

Comparing these genes more widely, it turns out that most animals are oriented like flies; it is our phylum, the Chordata, which is upside-down. Less is known about the evolution of the left–right axis, but we do at least know that two genes, *nodal* and *Pitx*, are involved in patterning this axis in animals as different as snails and humans.

The similarities are not restricted to the orientation of the body, but go right inside. For example, several of the genes controlling heart formation in vertebrates are also found in insects, where they too control development of a pulsating muscular tube. There are networks of genes specifying where an eye will form, and most of these genes are also the same whether the animal is a fly, a worm, or a human. Putting these remarkable findings together, it seems that the long-extinct ancestor of all bilaterian animals had a system of genes for telling dorsal from ventral, for distinguishing left from right, for telling cells where they are along the head-to-tail axis, and for building various internal structures and sense organs. These genes and their roles have been retained for hundreds of millions of years, albeit with some modifications, and with one of our ancestors turning upside-down for some reason. The French naturalist Étienne Geoffroy St Hilaire argued as much in 1830, but based on rather imaginative anatomical comparisons and a dubious idealistic basis. Geoffroy said 'There is, philosophically speaking, only a single animal.' His views were never accepted in his lifetime but, at least with regard to the bilaterians, it seems he may have been right.

The ancient set of genes used for building the body is sometimes referred to as the 'developmental toolkit'. Some of the toolkit genes, such as *Pitx* and the Hox genes, code for proteins that bind to DNA, switching batteries of other genes on or off. Others code for secreted proteins that transmit signals between cells, for example *nodal* and *dpp*, or interfere with signals, for example *sog*. These examples are just the tip of the iceberg, however, as the toolkit includes hundreds of genes coding for DNA-binding proteins, dozens producing secreted factors, and others coding for receptors to which the secreted factors bind. All can be found across diverse bilaterian phyla, though sometimes individual genes have been lost in the evolution of particular animal groups. The roles of the genes are often similar between different phyla, as in the examples above, but in other cases, toolkit genes have been recruited for different roles in divergent taxa. These are the ancient genes used to build the bilaterian body. But when did they arise? Does the evolution of the developmental toolkit tell us anything about the earliest steps in animal evolution?

Delving into the genome sequences of the non-bilaterian groups – sponges, placozoans, cnidarians, and comb jellies – reveals a rich picture. Some key genes are found in all animals, but several are not. Cnidarians, which might be the closest of the non-bilaterians to the bilaterians, have most of the toolkit genes, although their Hox cluster is less complex. The other phyla lack more of the toolkit genes; sponges lack Hox genes completely, for example. Stepping outside the animals, to the choanoflagellates, we see a bigger difference, with many of the toolkit genes absent. The conclusion is clear. The basic set of genes necessary to build animal bodies evolved around the time that multicellularity originated, but this set of genes was then expanded and elaborated through the earliest stages of animal evolution. A large toolkit of developmental genes was in place by the dawn of the Bilateria, half a billion years ago. Today, these genes are used to shape and pattern the myriad of animal bodies across all three great groups of bilaterians: the Lophotrochozoa, the Ecdysozoa, and the Deuterostomia.

Chapter 6
# Lophotrochozoa: wondrous worms

> It may be doubted whether there are many other animals which have played so important a part in the history of the world as have these lowly organized creatures.
>
> Charles Darwin, *The Formation of Vegetable Mould through the Action of Worms* (1881)

## Annelida: living ploughs and bloodsuckers

Just a year before his death, Charles Darwin published his last book. The work was enthusiastically received and, at least initially, sold even faster than had *The Origin of Species*. The unlikely bestseller, *The Formation of Vegetable Mould through the Action of Worms with Observations on their Habits*, included insights drawn from Darwin's own practical research, conducted on and off over 40 years. Now a grandfather and feeling his age, he was keen to publish his findings about earthworms 'before joining them', as he put it. The book's most important conclusion was that earthworms should not be despised as pests that left unsightly casts on well-manicured Victorian lawns, but rather they were 'living ploughs' crucial to the health of soil. Darwin showed that earthworms drag organic matter such as leaves underground, that their tunnels aerate the soil and provide channels for water drainage, and that their actions mix the soil preventing layers from becoming compacted, thereby promoting plant growth.

Earthworms even have effects on geology, through attrition of rocks and stones, and on archaeology by burying ancient remains.

Earthworms belong to the phylum Annelida and their anatomy is central to why these animals have such an impact. Annelids are soft-bodied, muscular, and elongate, with a mouth at one end and an anus at the other. They have a series of fluid-filled spaces, or coeloms, along the body, providing a degree of rigidity through internal water pressure, coupled with extreme flexibility. All these features are helpful for squeezing through spaces in soil, tunnelling, or even shifting soil by passage through the gut. But the one feature of annelids that is most crucial to the process is undoubtedly division of the body into a series of units or rings. It is a character obvious at a first glance and one that gives annelids their common name of 'segmented worms'. Splitting the body into segments, each with its own muscles, coeloms, and nervous control, allows earthworms to contract some parts of the body, making them long and thin, while at the same time other parts are squeezed lengthways to become short and fat. Thin parts probe forward into crevices, while fat parts anchor the worm still, and by passing waves of contraction from head to tail, the body is pushed forward through the soil.

There are more than 15,000 species of annelid worm, with most of these living in the sea and fresh water, rather than on land. Throughout their evolutionary diversification, segmentation has been key. For example, the predatory marine ragworms contract some segments on their left side and others on their right, contorting the body into moving side-to-side waves. This rapid but coordinated wriggling propels the animal forward at speed, enabling it to chase and catch its prey. Other marine annelids are more passive, living sedentary lives in burrows or tubes, filtering particles from the sea water. But even these worms deploy segmentation to great effect, as coordinated waves of contraction are used to flush out their burrow or tube and bring in fresh, oxygen-rich water.

One well-known group of annelids, however, has all but lost segmentation, and for good reason. This group is the Hirudinea, better known as leeches. Some leeches are predators, devouring small aquatic invertebrates. Others, as every tropical explorer or movie fan knows, feed by latching onto the flesh of larger animals and sucking their blood. These parasitic leeches attach to skin using a strong sucker, inject a powerful anticoagulant to stop blood clotting, and slice at flesh using three blade-like jaws. Since food sources such as horses, deer, fish, and human legs come along quite infrequently, leeches are adapted to take in giant meals when they get a chance, and one such modification affects segmentation. Leeches evolved from aquatic annelids not dissimilar to earthworms, but have lost the walls, or 'septa', that separate the segments internally. This allows the body to stretch as the leech gorges on blood, distending its body like a balloon. The downside of this modification is that leeches cannot manage the same coordinated waves of contraction used by earthworms or ragworms, and most species resort to less efficient looping movements.

The capacity of some leeches to drink human blood, rather painlessly, has been exploited by medicine for centuries. Two thousand years ago, the Greek physician Themison of Laodicea wrote of using leeches for bloodletting, and the practice continued in many parts of the world until well into the 19th century. Leeches provided a way to draw off 'bad blood' and correct 'imbalanced humours', seen as the source of many ailments for which the true cause was unknown. Even the word 'leech' derives from '*loece*', the Anglo-Saxon word for 'physician' or 'to heal'. The demand for leeches in the 18th and 19th centuries was huge, with natural populations of the large medicinal leech, *Hirudo medicinalis*, being so over-exploited that the species is still rare today across much of Europe. Leech farms sprang up, but even these could not keep pace with demand. This is hardly surprising, since by the 1830s, an astonishing 40 million leeches were being imported into France each year. But leeches are not confined to the history of medicine, and their use has made a surprising comeback in recent years. A common complication during modern microsurgery is venous insufficiency, which occurs when surgeons have been able to repair arteries but not the much smaller and thinner-walled veins, leading to build-up of blood pressure in the reconstructed or reattached tissue. This can be relieved by using leeches to drink some of the excess blood and to inject anticlotting agents, giving time for natural healing of the tissue. The technique has been used successfully during surgery to reattach or repair eyelids, ears, penises, fingers, and toes.

Three other groups of worms, now included within the annelids, each used to have phylum status until their probable evolutionary position was resolved by molecular analysis. These are the Echiura, or spoon worms; Sipuncula, or peanut worms; and the deep-sea Pogonophora, or beard worms. The former two groups are not segmented, and it is thought they have lost this character during evolution, in much the same way as leeches, although to a more extreme extent. The pogonophorans were also long thought to be unsegmented, until 1964 when some specimens dredged from the sea floor were found to have a short segmented tail, or 'opisthosoma'. It was suddenly realized that all previous descriptions were – rather embarrassingly – based on incomplete specimens. Many pogonophorans live in muddy burrows and are incredibly thin, elongate animals just a few centimetres in length. Others are giants, however, and build erect tubes stuck to rocks in the depths of the ocean.

The first large tube worms were discovered in 1969 by the United States Navy while operating a deep-sea submersible off the coast of Baja, California. Although these animals, *Lamellibrachia barhami*, were 60 to 70 centimetres in length, and dwarfed all previously known pogonophorans, it was only a few more years before some real giants were found.

The most famous discoveries came in the late 1970s when geologists using the deep-ocean submersible *Alvin* started to

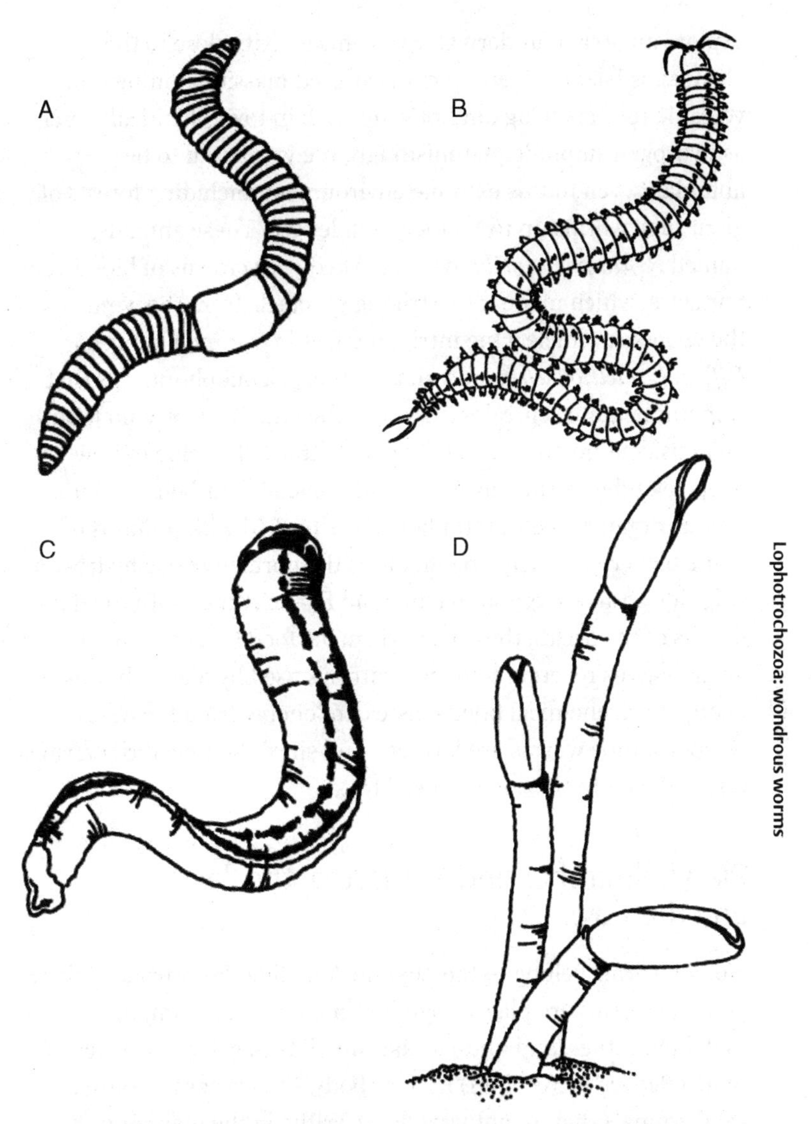


**6. Phylum Annelida: A, earthworm; B, ragworm; C, medicinal leech, *Hirudo*; D, pogonophoran, *Riftia***

explore an area of underwater volcanic activity close to the Galapagos Islands. Here were discovered massive chimneys of volcanic rock spewing out hot water rich in toxic chemicals, such as hydrogen sulphide. Astonishingly, life was found to be abundant even in this extreme environment, including forests of giant tube worms up to 1.5 metres in length. These animals, named *Riftia pachyptila*, were adorned with crowns of blood-red tentacles, which made for a striking spectacle from the windows of the deep-sea vehicle. One intriguing fact in common between *Riftia*, *Lamellibrachia*, and all the other pogonophorans, is that something quite central is missing – the gut. They have no mouth, no anus, and no obvious way to eat anything. The clue to how deep-sea tube worms survive is found inside their bodies. Here, a unique organ called the trophosome is packed with millions of living bacteria, all of a type that uses the normally toxic hydrogen sulphide as an energy source to build food chemicals. In the dark depths of the ocean, these bacteria make food by 'chemosynthesis': an analogous reaction to photosynthesis used by plants, but using energy from chemical bonds instead of energy from the Sun. Deep-sea tube worms are farmers, and since their bacterial farm is inside the body, they do not need to eat.



## Platyhelminthes and Nemertea: the flat and the slow

Not all worms belong to the phylum Annelida. Flatworms, flukes, and tapeworms are placed together in another phylum, the Platyhelminthes, and quite unlike annelids they have no trace of a fluid-filled cavity (coelom) in their body. Flatworms are rather solid animals that do not wriggle or writhe in the manner of annelids, because their muscle blocks are not split into individual segments and because the lack of a fluid skeleton means there is nothing rigid for the muscles to bend. Instead, flatworms move by sending small ripples of muscular contraction along their edges or, in the smallest species, by using cilia protruding from surface cells. Their lack of a circulating blood system or specialized gills means these animals rely on simple diffusion across the body surface to get oxygen to their cells, which in turn restricts most members of the group to a small size and flattened shape. Flatworms can easily be found by turning over small rocks in streams and rivers, where their oval bodies, a few millimetres to a centimetre in length, creep slowly and steadily along, grazing on algae and debris.

Not all platyhelminths live such innocuous lives, however, and several have unwelcome associations with humans. Perhaps the most important is the fluke *Schistosoma mansoni*, the agent of bilharzia, which currently infects over 200 million people. Symptoms of infection are variable, but in severe cases bilharzia can cause damage to internal organs and even death. Like many flukes, the bilharzia parasite needs two different hosts to complete its life cycle. After developing in a fresh-water snail, the schistosome larvae are released into the river where they will seek out and penetrate the skin of the second host, usually a human.

Another phylum of unsegmented worms is the Nemertea, or ribbon worms, commonly found on the underside of seashore rocks. These worms live life in the slow lane, creeping along at a leisurely pace and generally expending little energy. They too lack large fluid-filled cavities, enabling them to deform and stretch their bodies into contorted shapes, like stringy pieces of chewing gum. Despite their sluggish lifestyle, many ribbon worms are voracious predators that catch and eat other invertebrates using a long proboscis armed with either sticky glue or poisonous barbs. Most ribbon worms are only a few centimetres in length, although one British species, the bootlace worm *Lineus longissimus*, has a serious claim to be the longest animal on the planet. Specimens certainly reach 30 metres, about the same as a blue whale, but claims in excess of 50 metres have been made. Even in these monsters, however, the body of the worm is never more than a few millimetres wide.

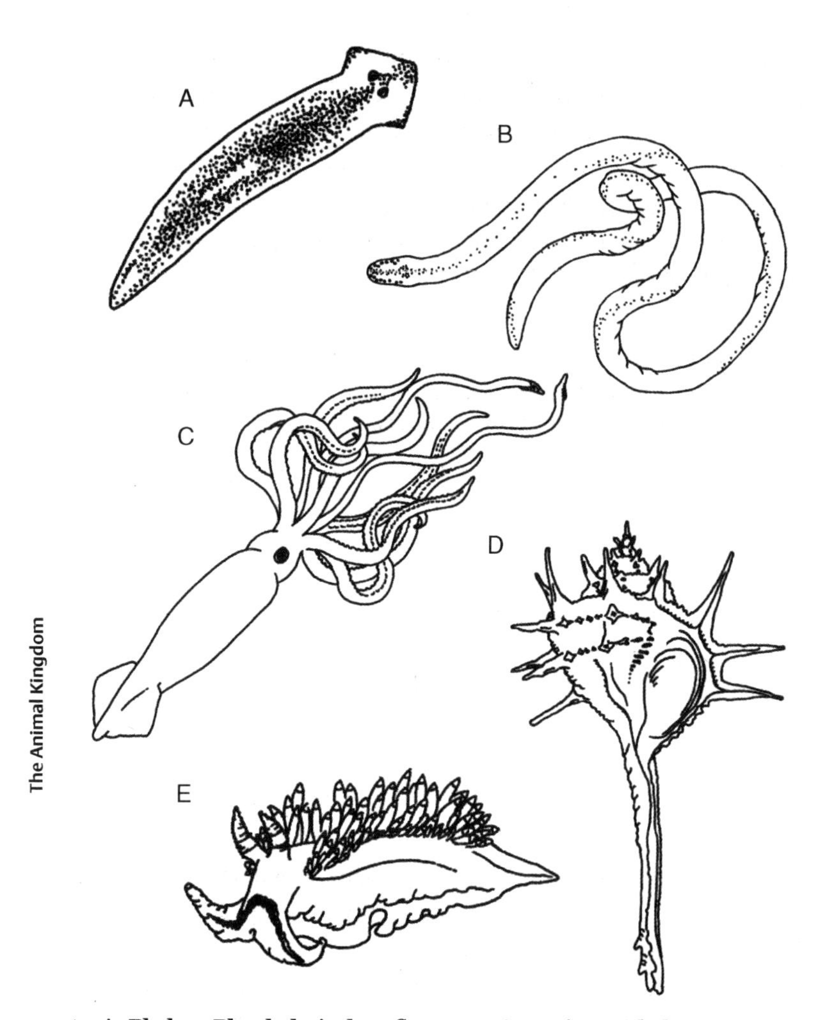


**7. A, Phylum Platyhelminthes, flatworm, *Dugesia*; B, Phylum Nemertea, ribbon worm, *Lineus ruber*; C–E, Phylum Mollusca: C, giant squid, *Architeuthis*; D, gastropod, *Murex brandaris*; E, aeolid nudibranch**

## Mollusca: from squid to snail

The honour of being the 'largest' invertebrate is usually given to a member of a different phylum: the Mollusca. The giant squid, *Architeuthis*, is an immense animal. It may not grow to anywhere near the length of the bootlace worm, reaching a 'mere' 13 metres, but on sheer bulk it wins hands down, or rather tentacles down. Like other squid, *Architeuthis* has a relatively short stumpy body, eight arms bearing suckers, and two additional tentacles with serrated suckers at the end. The animal lives in the ocean depths and, even though zoologist Tsunemi Kubodera recently succeeded in photographing and filming live giant squid, most of our knowledge still comes from specimens washed up onto land or accidentally caught by fishing trawlers. Alongside the ancient myths and legends of monstrous 'kraken', there is at least one plausible account of a giant squid attacking a ship, logged in the diaries of a Norwegian naval vessel, the *Brunswick*, from the 1930s. The French trimaran *Geronimo* also encountered a giant squid while competing for the Jules Verne Trophy in 2003, with one crew member describing its tentacles as being 'as thick as my arm'. Stories of squid attacking swimmers are more likely to involve a different animal, the Humboldt squid, *Dosidicus gigas*, which grows to a chunky 2 metres. These bold predators hunt in large shoals and can strike at fish, and other swimming prey, with rapidity and ferocity. No wonder some divers have taken to wearing body armour when swimming with packs of Humboldt squid. Along with octopus and cuttlefish, squid are cephalopods – one of the major groups within the Mollusca. Size is not their only claim to fame, and octopus in particular probably have the highest cognitive development of any invertebrate. Their brain is large and complex, their vision is acute, and they are able to solve puzzles such as mazes designed to test spatial memory.

Unlike most cephalopods, the majority of molluscs have a conspicuous shell. Secreted by a specialized layer of cells, the mantle, and composed of bricks of calcium carbonate, the main function of a shell is protection from predators. In gastropod molluscs, such as snails, the single shell is carried on the animal's back, with many of the delicate internal organs hidden away inside. Despite its obvious usefulness, some groups of gastropods have lost the shell completely in evolution. Many of these have found alternative means of protection. Terrestrial slugs, loathed by gardeners, secrete distasteful slime which deters some, though not all, predators. The lack of a shell is a positive advantage in some ways because, unlike snails, slugs can thrive in habitats with low calcium content.

One group of sea slugs, only distantly related to the land-living slugs and snails, has evolved an even more impressive means of defence. Aeolid nudibranchs feed on cnidarians such as sea anemones, but instead of being stung, they manage to harvest the nematocysts, or stinging organelles, without them firing. These subcellular structures are then recycled by the sea slugs and loaded into fronds of tissue growing from the upper surface of their body. Consequently, just like sea anemones and jellyfish, these marine gastropods bristle with a stolen armoury of explosive poison-tipped harpoons. Animals from the third major group of molluscs, the bivalves, have two shells. Oysters, clams, and mussels are well-known examples, and each has a similar way of life. Hidden between the two shells, the animal has elaborate W-shaped gills covered in thousands of cilia which beat to suck in a strong current of fresh, oxygen-rich water. The water flow also carries in tiny particles of suspended food such as microscopic algae that are swept towards the animal's mouth.



Molluscs have been an important food source for humans for millennia. Vast 'middens' or mounds of ancient discarded shells, often hundreds of metres in length, are found throughout coastal regions of the world. In addition to their use as food, both Pliny the Elder and Aristotle wrote of useful pigments that could be obtained from molluscs, particularly the dye Tyrian purple. Used to colour the robes of Greek and Roman noblemen, this vivid dye was extracted from the marine gastropod *Murex brandaris* by heating its bodily extracts after mixing with salt. Some species of mollusc impact on humans in more detrimental ways, such as *Biomphalaria*, the fresh-water snail encountered earlier as the intermediate host for the bilharzia parasite. One mollusc may have even changed the course of European history. In 1588, the

Spanish Armada sailed for England, determined to overthrow the monarch Queen Elizabeth I. The Spanish defeat is well recorded, but perhaps not all the credit should go to Sir Francis Drake. Before sailing for battle, the Spanish fleet anchored in Lisbon harbour for several months, where their wooden hulls became infested with the wood-boring bivalve *Toredo navalis*. This mollusc, the notorious 'shipworm', has an elongated body with its two shells reduced to small plates at one end, used for tunnelling into its food source – wood. With their timbers riddled with holes, the Armada was fatally weakened even before the battle began. The shipworm's habits are also the reason why so few historic sailing vessels remain today. The Swedish battle galleon *Vasa* is a beautifully preserved exception; *Vasa* sank on her maiden voyage in 1628 in the Baltic, a sea not salty enough for *Toredo* to live.

All the above phyla – Annelida, Platyhelminthes, Nemertea, and Mollusca – are part of the Lophotrochozoa: one giant arm of the evolutionary tree of animals with one giant name. The 'trocho' part of the name is derived from the 'trochophore', a particular type of planktonic larva possessed by some, though not all, species in these phyla (most clearly in marine annelids and molluscs). Trochophore larvae are usually described as resembling miniature spinning tops, although since trochophores do not actually spin, perhaps pear-shaped is a more useful description. The 'lopho' part of 'Lophotrochozoa' refers to the 'lophophore', an unusual feeding structure resembling a crown of arms, found in none of the phyla discussed above. Instead, lophophores are found in three additional phyla that are not particularly worm-like. These are the shelled Brachiopoda, the rare Phoronida and the minute Bryozoa, or moss animals, commonly found as mat-like colonies on fronds of large seaweeds. The best way to see a lophophore in action is to find a frond of kelp washed into a rock pool, and examine any white 'sea mats' under a low-power microscope or hand lens. Within a few minutes of being submersed, hundreds of tiny bryozoans will start wafting their delicate tentacles in the sea water, searching for particles of food. It took DNA sequence comparisons to reveal that the animals with lophophores fall in the same part of the Animal Kingdom as the animals with trochophores. This group, the Lophotrochozoa, is the 'sister' superphylum to another great grouping of bilaterian animals: the Ecdysozoa.

Chapter 7

# Ecdysozoa: insects and nematodes

To a good approximation, all species are insects.

Robert May, *Nature*, 324 (1986): 514–15

## Insects: masters of the land

Nobody knows how many species of insect exist. Estimates range from a few million to over 30 million. At least 800,000 different species have been described and named formally, but even this figure is not known accurately since no master list has ever been compiled. And for those species that have been described, the geographic distribution, ecology, and behaviour are unknown for a large proportion. But why are there so many species of insect? This is not a simple question to answer, but the reasons are likely to include a body plan that can be readily adapted to many ecological niches and food plants, coupled with the great diversity of plant species, particularly in the tropics. In addition, the land-living insects emerged from the sea early in animal evolution, almost 400 million years ago, giving time for tremendous diversification alongside the evolving land plants.

Insects are the supreme land animals. They are part of the phylum Arthropoda and, like all arthropods, insects have an external hard skeleton which is moulted intermittently to allow growth, plus a series of jointed limbs used for locomotion and feeding. Although their ancestors lived in the sea, the insects have evolved a suite of adaptations to permit life in a harsh environment prone to extremes of temperature and severely limited in water – the hostile environment we call land. An external skeleton provides inherent support to the body, whether on sea or on land, but in insects this cuticle has been waterproofed by addition of waxes into the outermost layer. This effectively stops desiccation caused by evaporation from the outer body surface. This solves part of the problem, but still leaves two processes prone to water loss. First, animals need to obtain oxygen and get rid of carbon dioxide, and the physics of gaseous diffusion dictates that this is most efficient across a wet surface. To avoid exposing wet surfaces to the outside world, which would defeat the point of a waterproof exoskeleton, insects have evolved an elaborate series of cuticle-lined 'trachea', or tubes which twist and branch from closeable pores on the outside of the body right into the inner tissues of the animal. Here the cuticle covering is absent and gas exchange can occur exactly where it is needed. Second, all animals need to dispose of nitrogen-containing waste products, which are generated during metabolism of proteins but which can be toxic to cells. Many animals, including humans, overcome this by diluting the waste products and excreting liquid urine, but this is wasteful of water. Insects mainly use a different metabolic pathway and generate uric acid, which crystallizes as a solid – in contrast to soluble ammonia or urea – and then use efficient glands to resorb water before excretion. Since uric acid is not toxic, many insects store some of the chemical in specialized cells, while others actually use it. The white colouration of *Pieris* butterflies, such as the Large White, is generated by uric acid storage in the wing scales.

An obvious characteristic shared by all members of the phylum Arthropoda is segmentation. Most of the major body parts, including muscles and nerves concerned with movement, are repeated in a serial manner along the length of the body, as if the body is divided into a set of units. This type of organization is similar to that of segmented worms, the Annelida, although –

contrary to long-held views – these two phyla are not at all closely related. Annelids fall within the Lophotrochozoa; arthropods are in the Ecdysozoa. In arthropods, the segmentation also affects the rigid external skeleton, so joints of softer cuticle are needed between segments to allow the body to twist and move. Without joints, the animal would be encased in an immovable coat of armour. In insects, the pattern of segmentation has been modified in a consistent way that may partly underpin the remarkable adaptability of insects. Instead of having a series of near-identical segments running the length of the body, groups of segments are fused together into three principal units, or 'tagmata'. First, there is the head, built of six or seven segments fused together, and containing the major nerve concentrations, sense organs, and jointed feeding structures. After a flexible neck joint comes the thorax, built of three segments fused together, each of which has a pair of jointed legs. In most insects, the second and third segments of the thorax, T2 and T3, each also have a pair of wings. And finally, the abdomen, made of eight to eleven segments fused together, albeit less rigidly, has no legs but encases the bulk of the digestive, reproductive, and excretory organs of the animal. In functional terms, the head is concerned with feeding and sensing, the thorax with movement, and the abdomen with metabolism and breeding. This separation of functions has allowed for optimization of each part of the body.



## Ruling the skies: wings and flight

Powered flight has evolved only four times in the history of life on Earth – in birds, bats, pterosaurs, and insects. The only flying invertebrates are insects. They are also the most abundant, and the most diverse, fliers on the planet. Flight is absolutely key to understanding the success of the insects. It is interesting, though a little puzzling, that insects evolved two pairs of wings. Are two pairs better than one? After all, birds and bats have only one pair of wings, although some dinosaurs, probably related to the ancestors of birds, had feathers on both 'arms' and 'legs'. The reason for the difference in numbers of wings might relate to a limitation imposed by the mode of embryonic development in vertebrates. There is some evidence that vertebrates, such as bats and birds, are constrained to have only two pairs of limbs; if one is needed for walking, then only one pair is left for flying. The wings of insects, in contrast, did not evolve from legs and no such constraint exists. So while segment T1 just has legs, segments T2 and T3 have both wings *and* legs. Having two pairs of wings allows for even greater diversity in the range of flying styles that insects can adopt.

The insects are divided into almost 30 'orders' including the grasshoppers (Orthoptera), dragonflies (Odonata), mayflies (Ephemeroptera), stick insects (Phasmida), earwigs (Dermaptera), cockroaches (Dictyoptera), land and water bugs (Hemiptera), and fleas (Siphonaptera). But without doubt, the 'big four' of the insect orders, accounting for over 80% of described species, are the beetles (Coleoptera), butterflies and moths (Lepidoptera), bees, wasps, and ants (Hymenoptera), and flies (Diptera). Each is phenomenally diverse, and each has adapted its wings in a different way.



Lepidoptera have two fully developed pairs of wings. In some moths, there are spines or projections that link fore- and hind-wings together, but in many Lepidoptera the wings can be moved and controlled independently in flight. Wing shape varies enormously, from the feathery projections of plume moths to the elongate blade-shaped wings of South American *Heliconius* butterflies and the broad gliding wings of swallowtail butterflies.

Moths and butterflies may look delicate and ephemeral, but some species are robust and long-lived. Monarch butterflies, *Danaus plexippus*, overwinter in mass roosts in central Mexico, before undertaking a collective migration across North America. Each individual will fly for hundreds of kilometres, and within just a few butterfly generations their descendants can reach as far north

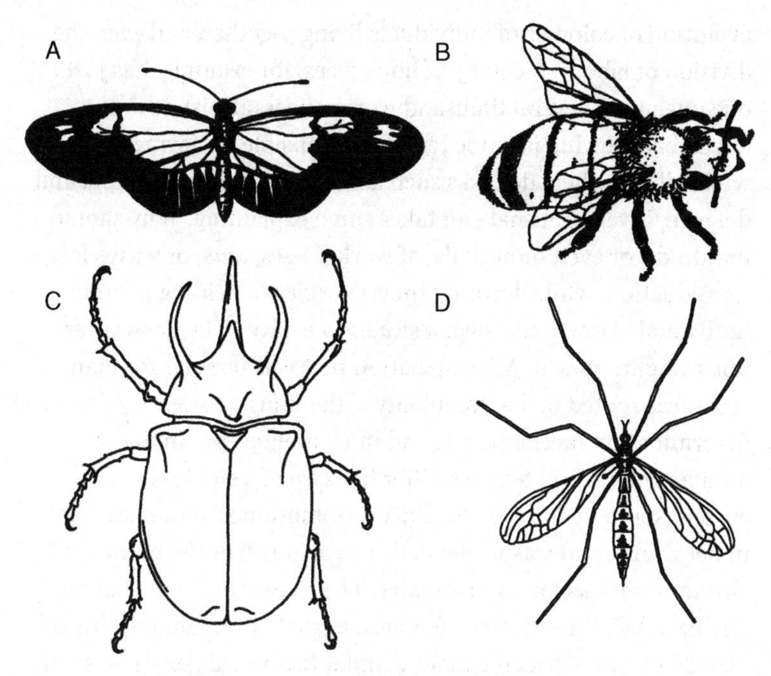


**8. Insects, the 'big four': A, Lepidoptera, butterfly, *Heliconius*; B, Hymenoptera, honeybee, *Apis*; C, Coleoptera, beetle, *Chalcosoma*; D, Diptera, crane fly, *Tipula***



as Canada – 4000 kilometres from the winter roosts. The Painted Lady butterfly, *Vanessa cardui*, is also famed for its migratory behaviour. Few European naturalists will forget the years 1996 and 2009 when vast swarms of Painted Ladies swept north across Europe from the Atlas Mountains of Africa, breeding as they went, to eventually reach as far north as Scotland and Finland.

The ants, bees, and wasps – Hymenoptera – also have two pairs of wings, but these are generally held tightly together by a row of hooks on the hind-wing. Most species are adapted for rapid, controlled flight, allowing bees to hover or dart into small spaces to collect nectar, hornets to catch prey in flight, and parasitic wasps to land near caterpillars into which they will lay their eggs. It is primarily within the Hymenoptera that we also see the evolution of colonies of individuals living together, and even the division of labour. A colony of honeybees, for example, has just one queen but several thousand worker bees, all sisters of the queen bee. Having just one female responsible for laying eggs, while all the others do tasks such as food gathering, cleaning, and defence, is very unusual and takes some explaining. Why should hundreds, or even thousands, of worker bees, ants, or wasps forgo reproduction, while devoting their energies to helping another individual? How could such a situation evolve? The answers are not straightforward. An explanation that was popular for many years was rooted in 'haplodiploidy' – the unusual sex determination mechanism found in Hymenoptera. In many animals, males and females differ because of a single sex chromosome, such as the X and Y chromosomes of humans. But in bees, ants, and wasps, the males have only half the number of chromosomes as found in females. This is because eggs that are fertilized by sperm become females; eggs that remain unfertilized, instead of dying, become males. Under this weird genetic system, sisters – such as worker bees and queen bees – are genetically very similar to each other. Indeed, a female ant, bee, or wasp is more related to her sisters than to her own children. This might imply that cooperation between sisters is favoured evolutionarily, because by helping the queen, the workers are incidentally promoting survival of their own genetic lineage. However, the catch with this often quoted explanation is that, in haplodiploidy, sisters are only weakly related to their brothers, cancelling out the genetic advantage. Instead, the evolutionary origins of sociality in ants, bees, and wasps may have less to do with unusual genetics, and more to do with shared defence of resources by relatives and a breeding system in which extended care of the young is the norm.

The linking together of fore- and hind-wings, as in Hymenoptera, means that two pairs have virtually the same mechanical properties as one pair. Two of the largest orders of insects have gone one step further and only use a single pair of wings for flight. Coleoptera, or beetles, fly using only their hind-wings. Diptera, the true flies, use only their fore-wings. Both groups certainly evolved from insects that used two pairs of wings for flight. In beetles, the ancestral fore-wings have evolved into hardened wing cases (elytra) that cover and protect the hind-wings when not in use. This modification has opened up new ecological niches for beetles; they can burrow into soil, bore into seeds, or tunnel in rotting wood without damaging their thin and delicate flying wings. The great diversity of beetles has fascinated generations of biologists, and the young Charles Darwin was a Coleoptera fanatic. In *The Descent of Man*, Darwin enthuses about one of the giant scarab beetles, writing:

> If we could imagine a male *Chalcosoma* with its polished, bronzed coat of mail, and vast complex horns, magnified to the size of a horse or even of a dog, it would be one of the most imposing animals in the world.



In the true flies, Diptera, the ancestral hind-wings have been modified into tiny club-shaped 'halteres'. These are vibrated up and down during flight, out of phase with the flapping of the fore-wings, and they form part of an intricate sensory feedback system. If the fly's body tilts to one side, the halteres are liable to continue their original plane of vibration, with the inertia of a gyroscope, and sense organs at the base of the haltere will then detect the change in angle between haltere and body position. The fly, therefore, receives continuous information about its precise orientation in space. Consequently, true flies are the most agile of all insects, able to hover, dart, or reverse with the most astonishing rapidity and accuracy.

Amongst the tens of thousands of dipteran species are several that have profound impacts on human life. These include mosquitoes that transmit the malaria parasite or carry the viruses causing yellow fever and dengue fever. Over one million people die from mosquito-borne diseases every year. Other flies have beneficial effects, for example as pollinators, while one species, *Drosophila melanogaster*, has played an important role in medical research. This species of fruitfly has been a favourite 'model organism' in genetics research for over a century, being small and easy to breed in large numbers. Research using *Drosophila* has given deep insights into gene functions and interactions, with relevance to many human conditions including cancer.

## Yet more arthropods: spiders, centipedes, and crustaceans

In addition to insects, there are three other classes of living animals within the phylum Arthropoda. Two of these, the chelicerates and myriapods, have successfully invaded land. The third, the crustaceans, is mainly aquatic but with some terrestrial members. Chelicerates include spiders and scorpions and, although most live on land, they had their origins in the sea. Their structure and adaptations to terrestrial life are so different to those of insects that it is clear the chelicerates and insects invaded land independently.



Turning to myriapods, the centipedes and millipedes are the best-known groups. These animals have a specialized head followed by a series of many segments bearing jointed legs. In centipedes, the body segments are separated by flexible rings of cuticle, allowing them to twist, turn, and run swiftly. Centipedes are predators that actively chase and catch their prey, attacking using ferocious 'poison claws' – giant venomous fangs that evolved from the front pair of legs. Millipedes, in contrast, eat mostly wood or decaying leaves, and are much slower animals. They lack poison claws, and in many species the segments interlock to allow millipedes to drive through soil or rotting vegetation like slow-motion battering rams. It is not true that centipedes have a hundred legs, or that millipedes have a thousand, although when it comes to numbers, there are some peculiarities that are not fully understood. Oddly, centipedes always have an uneven number of pairs of walking legs (not counting the poison claws), which means that while centipedes can have as few as 30 legs (= 2 x 15) or as many as 382 (= 2 x 191), no species has exactly 100. Even in species where segment number varies, individuals always differ by a multiple of two pairs. Millipedes have a different oddity. When viewed from above, millipedes seem to have two pairs of legs for each segment, leading to the idea that pairs of segments became stuck together in evolution to form 'diplosegments'.

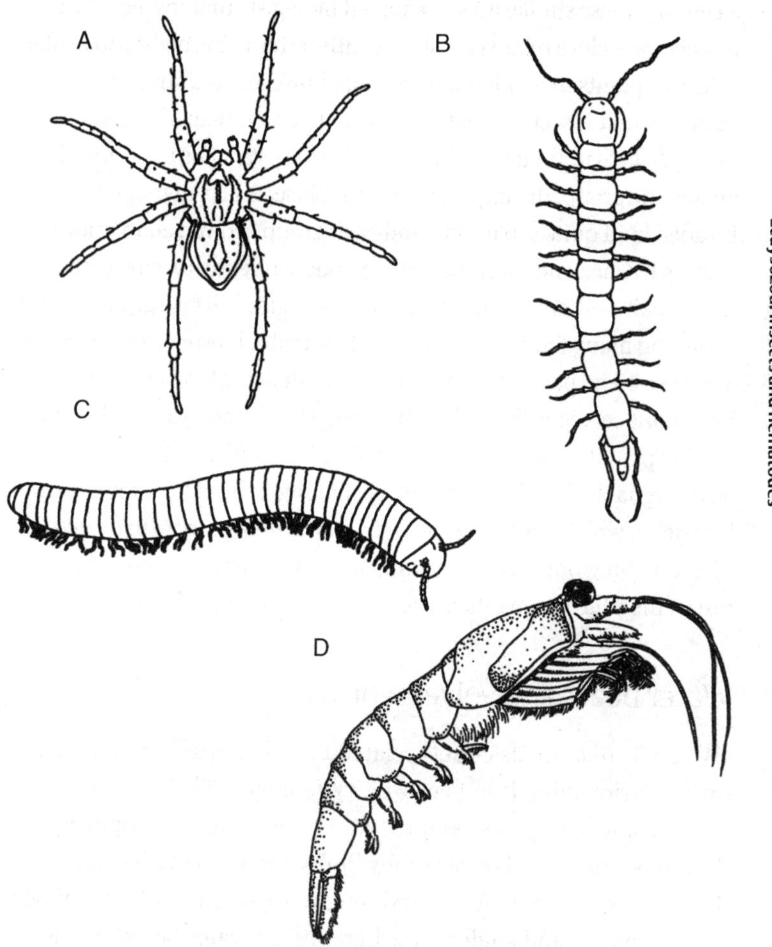


**9. Arthropoda: A, Chelicerata, spider; B, Myriapoda, centipede; C, Myriapoda, millipede; D, Crustacea, krill**

However, this pattern is not seen from below, and recent gene expression studies have revealed that the boundaries between segments are set independently on the top and bottom sides. In millipedes, segments are no longer simple repeating units.

Like insects, myriapods have trachea to deliver oxygen to their tissues and have limbs that do not branch. The structure of the head is also very similar between centipedes, millipedes, and insects. For over a century, these similarities persuaded biologists that myriapods and insects were close relatives – sisters within the arthropods. Molecular evidence points in a different direction, however, and strongly indicates that insects are actually closer to crustaceans. Indeed, insects probably lie within the crustaceans. Since crustaceans are primarily an aquatic group, the implication is that insects and myriapods invaded land quite separately, and each group evolved adaptations such as trachea and unbranched legs independently to cope with life in their new environment. Crustaceans are a diverse group, including many familiar animals such as crabs, lobsters, and shrimps, and even some parasitic species such as fish lice. Many are hugely important ecologically, such as the copepods found by the billion in marine plankton or the vast shoals of krill on which baleen whales feed. Arguably the most unusual, yet amongst the most familiar, are barnacles, which start their life as free-swimming larvae in the sea before settling onto rocks, sticking down head-first, and spending the rest of their life waving their legs to catch particles of food.

## Water bears and velvet worms

Two phyla of animals closely related to arthropods rank amongst the favourite animals of nearly every zoologist. These are the microscopic tardigrades and the forest-dwelling onychophorans. Both types of animal have stubby limbs and soft cuticles, rather than the rigid jointed limbs and tough exoskeletons of arthropods such as insects and spiders. Tardigrades, or 'water bears', are less than a millimetre in length, and can be found living in the surface water on damp moss or lichens.

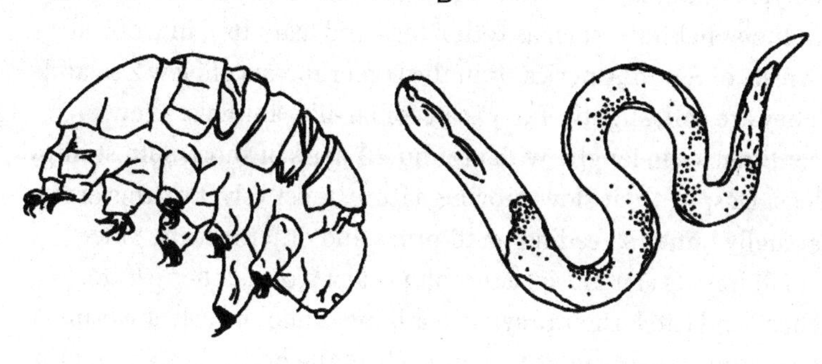


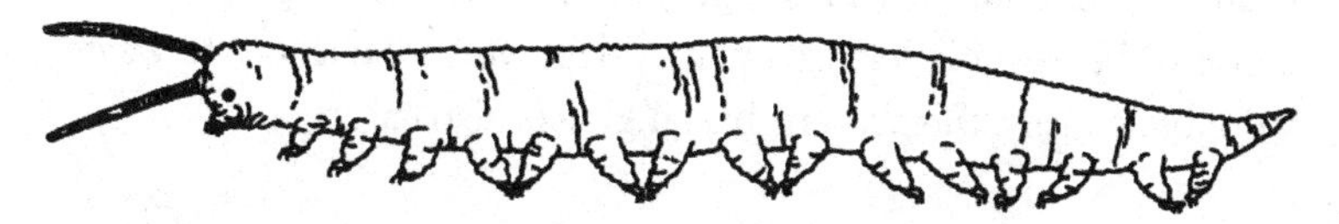


**10. A, Tardigrade, or water bear; B, Nematode, or roundworm; C, Onychophoran, or velvet worm**



Watched down a microscope, their chubby bodies and lolloping gait really do give them the appearance of miniature bears, albeit bears with eight legs. Apart from their general cuteness, tardigrades are famous for a remarkable ability to withstand extreme environmental conditions. If their habitat slowly dries up, tardigrades secrete a waxy covering and withdraw their legs until they resemble tiny barrels. They then vastly reduce their consumption of oxygen and water until they enter a state of almost suspended animation. In this condition, known as cryptobiosis (meaning 'hidden life'), tardigrades can survive for several years. Claims of a century have been made; however, such extensive survival times are unlikely in the light of recent studies. Tardigrades are also remarkably resilient, and once they have entered cryptobiosis, they have been known to survive temperatures as low as −200°C or as high as +150°C. Life is put on hold until conditions improve.

Onychophorans, or 'velvet worms', live on land, and can be found in moist habitats, such as rotten logs and leaf litter, in tropical forests of South America or in the cooler forests of New Zealand. They are soft, slightly 'furry', caterpillar-like animals, a few centimetres in length, with around 20 pairs of short, soft, stumpy legs. Despite their slow-moving nature, most velvet worms are actually hunters, feeding on termites and other insects. Since small insects can move faster than velvet worms, they cannot chase and catch their prey. Instead, they shoot it. Velvet worms have unusual appendages either side of the head, thought to have evolved from legs, which are used to fire streams of sticky slime at their intended targets. This glue entangles the insect prey, which can now be eaten at leisure by the velvet worm. So as not to waste valuable energy, the protein-based glue is also eaten.



## Moulting worms

The nematodes, or roundworms, are the most unlikely relatives of the arthropods. They are not segmented, they have no external skeleton, and they have no limbs. As their name suggests, they are simply worms – long, thin, and flexible. Yet, since 1997, more and more DNA sequence evidence has accumulated to indicate that nematodes do indeed sit quite close in the Animal Kingdom to the arthropods, water bears, and velvet worms, plus a few obscure animals such as the wonderfully named (but microscopic) kinorhynchs, or 'mud dragons'. Most zoologists were very surprised by the finding, which had never been suggested from studies of anatomy. But in fact, all these animals have one fundamental character in common – they shed their skins as they grow. Arthropods have a hard external skeleton, which must be shed repeatedly to allow the body beneath to enlarge, a process called ecdysis. Water bears and velvet worms (as well as immature insects such as caterpillars) have softer, more flexible cuticles, but these are also moulted because the molecular structure of their cuticle is not well suited to expansion. Nematode worms have complex cuticles built from tightly wrapped fibres of protein, winding round the body to form layers upon layers of densely packed elastic springs. These too must be shed to allow growth. When DNA evidence indicated that these animal phyla were related, a name had to be invented for the group. Anna Marie Aguinaldo, James Lake, and colleagues, the biologists who first spotted the relationship, named the group Ecdysozoa meaning 'moulting animals'.

Nematodes have a very unusual internal structure. They have a fluid-filled space in their body, as do many other worms, but they keep this fluid at very high pressure, around ten times the pressure of fluid in other worms. The internal pressure pushes out against the nematode's tissues and cuticle, giving them a circular cross-section. This explains their common name of 'roundworms'. Another peculiarity of nematodes is that all their body muscles run in the head-to-tail orientation (longitudinal), with no muscles wrapping around the body in circles. Most other worms, including earthworms, ragworms, and ribbon worms, have both types of muscle, which can contract in opposition to each other, or antagonize, allowing the animal to change shape and crawl or burrow. So how do nematodes manage to wriggle and move, if their muscles can only contract lengthways? The answer lies in the high-pressure fluid cavity and the elastic springy cuticle, which counteract the muscles, enabling the worm to throw its body into rapid undulating waves. Their movement is not so well coordinated as in earthworms or ragworms, partly because of the unusual muscle arrangement, but also because roundworms are not segmented and cannot easily make different parts of the body move in opposite directions. Instead, nematodes move with a thrashing motion, which is not very effective in swimming but works perfectly well in their preferred home – inside things. Many species of nematode live in soil or rotting vegetation; rotting fruit can be swarming with them. There is even a yeast-eating 'beermat nematode', *Panagrellus redivivus*. Many others are parasitic inside plants or other animals. Nor are humans immune from their attention, with some serious medical conditions caused by parasitic nematodes, including river blindness, guinea worm disease, toxocariasis, and elephantiasis.

The propensity of nematodes to live inside other organisms is described in a poetic, but somewhat exaggerated, 1914 quote from the 'father of nematology', Nathan Augustus Cobb:

> If all the matter in the universe except the nematodes were swept away, our world would still be dimly recognizable, and if, as disembodied spirits, we could then investigate it, we should find its mountains, hills, vales, rivers, lakes and oceans represented by a thin film of nematodes. The location of towns would be decipherable, since for every massing of human beings there would be a corresponding massing of certain nematodes.



Closely related to the nematodes, and with many similarities, are a group of exceedingly long and thin worms given their own phylum, the Nematomorpha. Although rarely more than a millimetre in width, these animals often reach 50 to 100 centimetres in length. Like nematodes, the nematomorphs have a tough cuticle, which is moulted as they grow, and only longitudinal muscle. Unlike nematodes, they do not eat anything. Or at least the adults do not, and the gut is shrivelled to a mere remnant. Juvenile nematomorphs certainly do feed, eating from the inside the body tissues of their arthropod host, which may be a grasshopper, cockroach, or freshwater shrimp. There, the worm will grow and moult, ever lengthening, until it reaches a size too great for the host animal, whereupon it bursts or crawls out, leaving behind the carcass of the unfortunate host. The adult worms must live in water, and so if the host is a land-living animal such as a cockroach, the parasite somehow manipulates the behaviour of the host, persuading it to move towards water and dive in to await its grisly death. The nematomorphs that parasitize water-living hosts, such as freshwater shrimps, gave rise to the common name for these animals: 'horsehair worms'. Long before the true life cycle of these animals was known, country people would occasionally notice long, thin worms swimming in apparently clean horses' drinking troughs, when no such creatures were present the day before. The myth arose that these were hairs from horses' tails that had come to life. The truth is less miraculous but more macabre: the giant worms were parasites that had burst out of tiny shrimps living unobtrusively in the water.

# Chapter 8
# Deuterostomes I: starfish, sea squirts, and amphioxus

I also here salute the echinoderms as a noble group especially designed to puzzle the zoologist.

Libbie Hyman, *The Invertebrates IV* (1955)

## Clues from embryos

Echinoderms have been called the strangest animals on Earth – and with good reason. Starfish, sea urchins, brittlestars, sea cucumbers, and sea lilies – the five classes within the echinoderm phylum – have much in common with each other, but little with anything else. They are built like nothing else on the planet. Even so, it has long been realized that they sit in the same part of the Animal Kingdom as do you and I: the deuterostomes. And the first clue to that relationship is to be found in embryos.

Most animals begin their life as a single cell, the fertilized egg. This divides to give two cells, then four cells, then eight, then sixteen, and so on. Although this seems straightforward, several different patterns can be seen if different bilaterian animals are compared. Two of the commonest are spiral cleavage and radial cleavage. The differences between them are quite clear if developing embryos are watched down a microscope. In spiral cleavage, when four cells divide to make eight, the new cells end up sitting above the grooves between the four old cells. If you tried to stack four oranges on top of four other oranges, this is exactly the pattern you would opt for. But in radial cleavage, the new cells sit directly on top of the four old cells, in a manner that would demand considerable balancing skills if attempted with oranges.

Whether spiral or radial, the same pattern is repeated in each subsequent cell division until eventually a hollow ball of cells is formed. Starting at a point or slit on the ball's surface, some cells then move inwards, rather like a finger or hand being pushed into an inflated balloon. The indentation where the cell sheet folds inwards is called the blastopore, the original ball of cells being the blastula. As the embryo develops further, this indented tube eventually forms the gut. And it is during this process that a second significant difference is found between the two patterns. In animals with spiral cleavage, the blastopore may mark the 'mouth' end of the gut, or more usually the blastopore is slit-shaped and closes up in the middle to leave two open ends: mouth and anus. But in animals with radial cleavage, the blastopore marks the rear end of the embryo, where just the anus will form. The mouth must break through quite separately at the other end of the developing embryo, as the incipient gut tunnels through to the far side. For this reason, animals with spiral cleavage have long been called 'protostomes', meaning 'first mouth' and alluding to the observation that the mouth develops from the first opening to form in the embryo. Animals with radial cleavage and blastopore at the rear were 'deuterostomes', meaning 'second mouth'. Inevitably, not all animals fit into either of these neat patterns, particularly if their embryos are endowed with lots of yolk which affects how the cells divide.

The distinctions were made by Karl Grobben in 1908, but a century later they need to be used with caution. Two of the great superphyla of bilaterian animals defined by molecular analyses, the Lophotrochozoa and the Ecdysozoa, include all the animals with the protostome mode of development, but they also include many other animals. For example, ecdysozoans such as insects and nematodes do not have spiral cleavage, nor indeed is it radial. Even so, 'protostome' is still often used today as a term to mean the Lophotrochozoa plus Ecdysozoa, with the knowledge that it is just a handy name not a unifying rule. Just as confusingly, the evolutionary grouping now called the Deuterostomia, and again defined by molecular analyses, is slightly different from Grobben's original. The group as now defined includes only some, but not all, animals with radial cleavage and secondary mouth formation. Perhaps it would have been better to discard the old names, but nomenclature is not always logical. Instead, simply note that not all protostomes have protostomous development, and indeed some animals placed in the protostomes actually have deuterostomous development. Consequently, not all animals with deuterostomous development are 'true' deuterostomes. There are only three major animal phyla in the Deuterostomia, as defined today, plus possibly one or two 'minor' phyla. The three major groups are the Echinodermata, the Hemichordata, and the Chordata.



## Life with the number five

Cut an apple open across its middle, and you will see a five-pointed star containing seeds. Take a close look at a wild rose and notice its five petals. Whether looking at fruit, flowers, or even leaf patterns, the number five is pervasive across the Plant Kingdom, and is the basis for much variation and adaptation. Animals, in contrast, despise the number five. One might point out that we have five fingers, but since we have two hands, the true number, of course, is actually ten (or twenty, if we count all digits). Five does not lend itself well to animals that have a central plane of symmetry, with a left- and right-hand side, as seen throughout most of the Animal Kingdom. Numbers such as two, four, six, and eight are everywhere, but not five. The basal animals, such as jellyfish in the phylum Cnidaria, do not have an obvious plane of left–right symmetry, but even these animals usually have four-fold symmetry, not five-fold.

Echinoderms do things differently. The evolution of the whole phylum has been dominated by the number five. The pattern is most easily seen in the starfish and brittlestars, common invertebrates of the sea shore and subtidal zone, which have five arms radiating from a central region or disk. In starfish, also called sea stars, the arms are held out quite rigidly. When the animal moves, it seems to glide over the sea bed, a trick achieved by the use of thousands of tiny 'tube-feet' projecting from the underside. The movement of the tube-feet is driven by an extensive series of fluid-filled canals in the body, unique to echinoderms, called the water vascular system. Although superficially similar to starfish, the brittlestars are different in that their five arms are thinner and more flexible, and used to assist the animal's locomotion by grasping and pulling. The two groups are also quite different ecologically, particularly from the viewpoint of a scallop. Brittlestars graze on debris and detritus, ingesting tiny pieces through a small downward-pointing mouth located in the middle of the central disk. Most starfish, on the other hand, are voracious predators. They may move slowly, but this does not matter if your prey doesn't move at all. Many starfish hunt bivalve molluscs such as mussels, oysters, and clams, animals that live sedentary lives inside their two tight-fitting shells. While generally protected from predators, starfish are the bivalves' worst nightmare. When a hunting starfish encounters prey such as a clam, it wraps its arms around it, holding tight using the sucker-like tube-feet, and pulls. As a tiny gap appears between the shells, the starfish then (rather riskily) pushes part of its own stomach inside-out, through its own mouth, and presses it into the gap. The stomach secretes enzymes to break down protein, weakening the clam's own muscles, so that the shells can be prised further apart. Eventually, the body of the clam is exposed and devoured. No wonder that those few species of bivalve that can swim, such as Queen scallops, make a break for freedom at the merest smell of a starfish.

The sea urchins, which bristle with defensive spines, and the soft, elongated sea cucumbers, are also echinoderms. Here, the

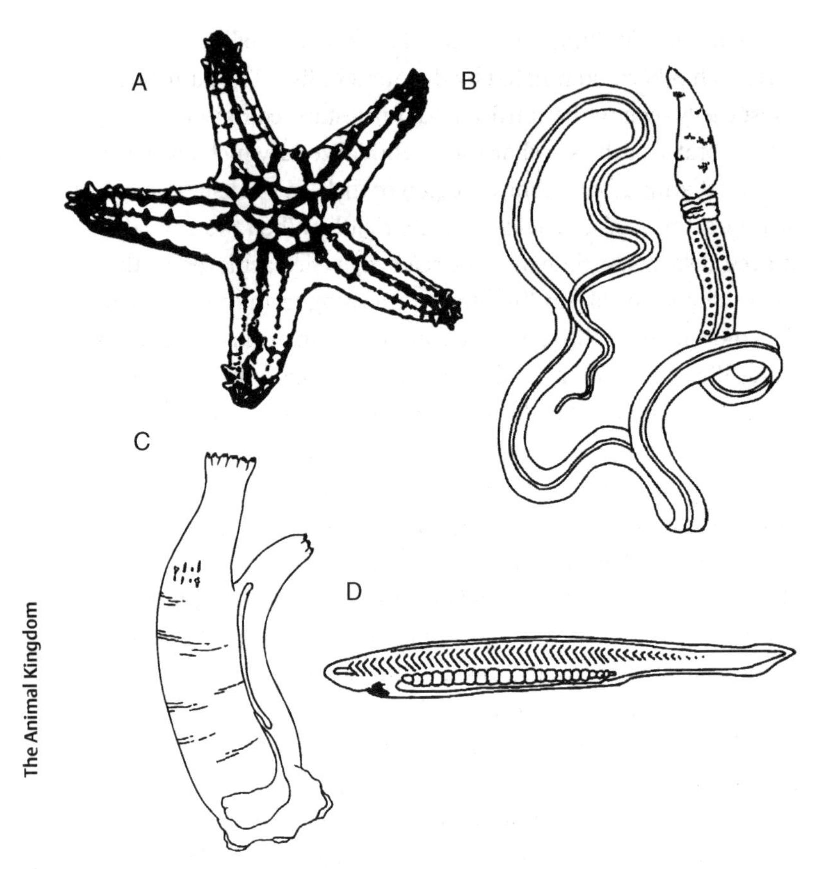


**11. A, Phylum Echinodermata, starfish; B, Phylum Hemichordata, acorn worm; C–D, Phylum Chordata: C, adult ascidian, *Ciona*; D, amphioxus, *Branchiostoma***

number five is less obvious to a casual glance, but it is certainly there. In each case, there are five zones around the body bearing tube-feet: an indication that these animals evolved from starfish-like ancestors in which the arms were folded up over the rest of the body. The fifth group in the phylum, crinoids or sea lilies, comprises filter-feeding animals with a mouth on the upper side, surrounded by a crown of five feathery arms. This sometimes sits atop a stalk, particularly in deep-sea species. With their tube-feet and upward-facing mouth, crinoids are oriented upside-down in comparison to starfish and brittlestars.

The evolutionary origin of five-fold, or pentaradial, symmetry in echinoderms is intriguing. It is clear that pentaradial symmetry evolved from bilateral (left–right) symmetry, for three key reasons. First, echinoderm larvae are bilateral, just like the larvae of many other marine animals. It is only when they settle out of the plankton and undergo metamorphosis that the five-fold pattern emerges. Second, echinoderm fossils have been found with all manner of symmetries, including bilateral, suggesting that the five-fold pattern was hit upon rather late in their evolution. Third, and most important, in the evolutionary tree of animals, the phylum Echinodermata is nested well within the Bilateria, indicating evolution from the same common ancestor as shared with all the bilateral animals alive today.

## Hemichordates: stinking worms

Some years ago, I wanted to collect an acorn worm. These are rather curious, non-segmented worms belonging to the Hemichordata, a phylum close to echinoderms on the evolutionary tree. Their early embryonic development is very similar to that of echinoderms, with radial cleavage and 'secondary' mouth formation, and their larvae – found occasionally in plankton samples – could easily be confused for echinoderm larvae. I had never seen an acorn worm in the wild, and I needed a sample for a research project. But on writing to marine biologists, I received one reply that baffled me. The writer told me that he too had never seen one in Britain, but he was absolutely certain they existed on a particular beach because he was sure he had smelt one. Of course, I was never going to believe the evidence of a simple aroma! That is, until I collected them myself. Most hemichordates are just a few centimetres in length, and live hidden away in sandy or muddy burrows where they filter food particles from the overlying sea water. They do this using a system of particle-trapping slits in their throat (pharyngeal slits) across which water passes, in a similar way to water passing over the gills of a fish. Indeed, the system is probably homologous, meaning that the shared ancestor of hemichordates and vertebrates had slits in the throat, used for acquiring food or oxygen. Many acorn worms do indeed have a striking medicinal smell, rather like iodine, which has been traced to a toxic chemical – 2,6-dibromophenol – found at high concentrations in their skin. The function of the substance is not fully clear, but it may deter predators intent on devouring them or it may limit bacterial growth in the burrow, or indeed both. Regardless of its adaptive function, the smell lingers on clothes and fingers, and once encountered is never forgotten.



The acorn worms are not the only members of the phylum Hemichordata. Their evolutionary sisters are a group of animals called pterobranchs: minuscule tube-dwelling animals with a crown of tentacles. They are not often encountered unless you know exactly where to look. The best-known British species, *Rhabdopleura compacta*, is less than a millimetre in length, and its tiny white tubes are found predominantly on the inner surface of the discarded shells of one species of mollusc, the dog cockle *Glycymeris glycymeris* – and even then, only in a few selected locations around the British coast. Other species can be found in Bermuda and in the fjords of Scandinavia, but their biology is still not well known. A distinct pterobranch genus, *Cephalodiscus*, was discovered on the seabed of the Magellan Straits by the famous *HMS Challenger* expedition in 1876, and when it was realized that this animal had pharyngeal slits, it became evident that pterobranchs were related to acorn worms, in the Hemichordata. A third pterobranch genus, *Atubaria*, is unusual in that it does not seem to live in a tube. Little is known about the biology of *Atubaria* because just 43 specimens have ever been seen, these all having been collected on 19 August 1935 by a marine expedition from the Imperial Palace of Japan.

## Tunicates: was man once a leather bottle?

Pull up the mooring ropes attached to buoys in a sea-water harbour and you will probably find them encrusted with hundreds of bottle-shaped leathery lumps, often in shades of yellow or brown, each a few centimetres in length. Pull one off its submerged home and it may squirt you in the eye. These animals, though they hardly look like animals, are sea squirts, or ascidians. Despite their amorphous appearance, these lumps are among our evolutionary cousins, members of our own phylum: the Chordata. Externally, sea squirts are encased in a tough outer covering or tunic which to the touch feels more plant-like than animal-like. This is because, remarkably, the tunic contains cellulose, a chemical usually found in plants not animals. There are two tubes, or 'siphons', on the top of the body, and sea water is sucked in through one and expelled through the other, the current being driven by the wafting of thousands of tiny cilia within the animal. This constant flow of water brings microscopic food particles and dissolved oxygen, and removes waste products.

A phylum is supposed to comprise a group of evolutionarily related animals with similar body layout. To repeat Valentine's words, 'phyla are morphologically-based branches of the tree of life'. So how could a sea squirt be in the same phylum as the vertebrates, along with you and me, along with birds and fishes? Looking at an adult sea squirt – the stationary, filter-feeding, cellulose-coated lump – there is little to suggest a close evolutionary relationship. And indeed, the earliest naturalists were completely unaware of the link. Aristotle considered sea squirts to be molluscs, like clams and snails, but he did note they were unusual in that their 'shell', actually the tunic, was leathery rather than hard and enveloped the whole animal. In the early 19th century, Lamarck removed them from the molluscs and erected a new group, the Tunicata, but he did not solve their affinities. All changed in 1866 when the brilliant Russian zoologist

Alexander Kowalevsky published a careful description of embryonic and larval development of a sea squirt, and moreover realized the deep significance of what he had found. The embryos of sea squirts develop into miniature 'tadpoles', usually just a millimetre in length, that swim in the sea for a day or two before landing head-down on a rock or other substrate. There they undergo a dramatic metamorphosis into a miniature version of the adult. From that moment on, the animal never moves place again and remains stuck to its landing spot, filtering sea water. Kowalevsky found that in the swimming tadpole stage, there is a small brain at the front, connected to a nerve cord along the back, which in turn lies on top of a stiffening rod, the notochord. These were all features characteristic of vertebrates, such as humans and fish, or at least characteristic of their embryos. An evolutionary relationship to vertebrates was clear.



News of the discovery swept through scientific circles, since there had already been much debate about which groups of invertebrates might be closely related to the vertebrates. In *The Descent of Man*, published in 1871, Darwin wrote:

> Some observations lately made by M. Kowalevsky, since confirmed by Prof. Kuppfer, will form a discovery of extraordinary interest.... The discovery is that the larvae of Ascidians are related to the Vertebrata, in their manner of development, in the relative position of the nervous system, and in possessing a structure closely like the chorda dorsalis of vertebrate animals. It thus appears, if we may rely on embryology, which has always proved the safest guide in classification, that we have at last gained a clue to the source whence the Vertebrata have been derived.

The emerging view, shared by Darwin, was that the long-extinct common ancestor from which sea squirts and vertebrates both evolved must have been a small tadpole-like animal, with the various features seen today in a sea squirt larva. But several other zoologists attempted to derive vertebrates from an ancestor much more like a modern sea squirt, complete with metamorphosis, and this view pervaded until late in the 20th century. Charles Neaves, a Victorian lawyer and poet who wrote much about evolution, beer, and women's rights, espoused this latter view in rhyme:

> How many wondrous things there be, of which we can't the reason see!
> And this is one I used to think, that most men like a drop of drink.
> But here comes Darwin with his plan, and shows the true Descent of Man:
> And that explains it all full well, for man-was-once – a leather bottèl!

Neaves may have found himself a reason to enjoy a drink (which he expanded on in eight subsequent verses), but he was not accurately reflecting Darwin's view or Kowalevsky's discovery. There is no need to assume that the common ancestor of sea squirts and vertebrates, our distant ancestor, had a life cycle that underwent metamorphosis like a modern sea squirt. Indeed, there are living relatives of sea squirts, called larvaceans, which never undergo the change and remain as swimming tadpoles all their life, feeding and reproducing.

## Amphioxus: the riddle of the sands

The phylum Chordata, or chordates, can be split into three evolutionary groups, or subphyla. In addition to the tunicates (such as sea squirts and larvaceans) and the vertebrates, there is a fascinating group of marine chordates called cephalochordates, more usually referred to as amphioxus or lancelets. Several characteristic features typify all chordates. These are a brain, a nerve cord running along the back rather than the belly, a notochord, repeated blocks of muscle on each side of the body, and 'pharyngeal' slits or holes between the throat and the outside world. These characters describe the typical body plan of a chordate. In the sea squirts, we saw most of these characters in the tadpole larva, except for the pharyngeal slits; it is the adult sea squirt that has these filtering structures. All the chordate features are seen in fish, with gills developing from pharyngeal slits, but the notochord becomes surrounded by bone during development and rather squeezed out of existence. As humans, we also have most of these characters at some stage in our development, but again the notochord is really only evident in the embryo, while our pharyngeal slits are just grooves in the embryo that never break through as actual holes. But the animal in which all the chordate features are most evident, even in the adult, is amphioxus. It has the clearest example of the chordate body plan one could ever hope to see.

There are about 30 species of amphioxus found in marine habitats around the world, often in tropical or subtropical seas, but sometimes in colder waters. One species lives off the European coast, and can be found buried in gravel in parts of the Mediterranean Sea and in the English Channel close to the treacherous Eddystone Reef with its famous lighthouse. Another species is common in subtidal sands around the Gulf Coast of Florida, while a third species was once so abundant near the city of Xiamen in China that there used to be a commercial fishery harvesting the animals for food. All species are roughly fish-like in overall appearance, just a few centimetres in length, with segmentally repeated muscles on either side of a prominent notochord acting as a stiffening rod. The springy notochord acts as an antagonist to the contracting muscles; this allows the animals to swim very rapidly when needed, for example when emerging from the sands to shed eggs or sperm into the sea water. The pharyngeal slits are very obvious and used for filtering algae from a current of sea water drawn in through the mouth. Unlike true fish, which are vertebrates, there is no bone, no fins emerging from the sides of the body, and a far less complex head. Amphioxus has the basic 'chassis' of a chordate, without many of the complications that have evolved in vertebrates.

A century ago, amphioxus was one of the most popular topics of research in all of zoology. In 1911, the great German evolutionary biologist Ernst Haeckel wrote that amphioxus was 'after man the most important and interesting of all animals'. I am tempted to agree with him. Even so, for Haeckel and his contemporaries, it posed a conundrum. On the one hand, many zoologists thought amphioxus must be a degenerate vertebrate, simply a fish that had lost many specialized characters. Others thought this highly unlikely, with Edwin Stephen Goodrich, Britain's greatest comparative anatomist, calling such a suggestion 'ridiculous'. Goodrich's view, backed up by careful studies of the animal's development and anatomy, was that amphioxus retained a more primitive chordate organization, not so different from the long-extinct ancestral chordate. This opinion was eventually accepted, and recently gained even stronger support from genome sequencing. Amphioxus, therefore, is another crucial link between invertebrates and vertebrates, rather like the position accorded to the sea squirt larva. In amphioxus, we have an animal, still living today, that has most vertebrate features in rudimentary form. It does have some of its own specializations, not least in its strange one-eyed head, akin to the cyclops of Greek mythology. These small changes occurred during the half a billion years since amphioxus and vertebrates last shared a common ancestor; exactly the same time frame that saw the emergence and diversification of the fish, amphibians, reptiles, birds, and mammals. So while amphioxus is not an ancestor of any other living animal, it seems to have changed remarkably little from the long-extinct ancestor of all the vertebrates.

# Chapter 9
# Deuterostomes II: the rise of vertebrates

The Romans in the height of their glory have made fish the mistress of all their entertainments; they have had Musick to usher in their Sturgeons, Lampreys and Mullets.

Isaak Walton, *The Compleat Angler* (1653)

## The great divide?

It is common to see textbooks of zoology concerned solely with the invertebrate animals, and other books devoted to the vertebrates (animals with backbones). Many university courses divide teaching of animal diversity along the same lines. And it is not a new division. Jean-Baptiste Lamarck, chiefly remembered today for his discredited ideas on the inheritance of acquired characters, made the distinction clear and wrote of '*des animaux sans vertebres*' two hundred years ago. Even he was not the first to draw this line, since over two thousand years ago, Aristotle divided the animals into 'sanguineous' (those with blood) and 'non-sanguineous' (those without); this was essentially the same vertebrate/invertebrate divide.

Despite its persistence and popularity, many zoologists have pointed to a deep-rooted problem with this view. The majority of individual animals are invertebrates, and most of the described animal species are also invertebrates. The difference in numbers is vast, with millions of invertebrate species and only around 50,000 different vertebrates. But the problem runs deeper than simply inequality. The problem lies in the evolutionary tree of animals, the history of animal life. Animals are classified into phyla, which represent branches on the evolutionary tree containing species with similar body organization. Of the 33 or so animal phyla, 32 are purely invertebrate. Even the 33rd is not a phylum solely of vertebrates, but contains a mix of invertebrate and vertebrate animals. It is our phylum, of course, the Chordata, and contains the invertebrate tunicates, the invertebrate amphioxus, and the vertebrates. The organization of the body is sufficiently similar in all these animals that they are grouped together. So taking a step back and looking at the diversity of the Animal Kingdom, vertebrates are not considered sufficiently different to even warrant their own phylum. Does this mean that vertebrates are but a twig in the tree of animal life?

## Differences run deep

While the numbers argument and the phylogenetic problem cannot be doubted, vertebrates do have some important specializations. Indeed, in some ways, they are exceptional animals. Most obviously, almost all the 'large' animals on the planet are vertebrates. There are a few big invertebrates, such as squid, octopus, and Goliath beetles, but the vast majority of invertebrates are less than a few centimetres in length. Contrast that to vertebrates, where large size is the rule. Fish, amphibians, reptiles, birds, and mammals – all have their giants. The 12-metre whale shark, 1.5-metre giant salamanders, 30-metre dinosaurs (extinct of course), the 3-metre-tall elephant bird (also sadly extinct), and the 30-metre blue whale may be record-holders, but each is simply at the end of a continuum. In fact, very small size is almost unheard of amongst the vertebrates. One of the most diminutive is an Indonesian fish, *Paedocypris*, with adults under a centimetre. Even this is a monster compared to many invertebrates.

One key to growing larger is rooted in the sophisticated system of veins and arteries used to deliver oxygen and remove carbon dioxide from active tissues deep inside the body: the highly efficient 'closed blood system'. Incidentally, some of the largest invertebrates, squid and octopus, also have a closed circulatory system, although this evolved independently. A second and equally important character permitting large size is the one that gives the vertebrates their name: the vertebral column. Skeletons come in many forms around the Animal Kingdom. Many worms have fluid-based support systems, arthropods have tough external skeletons, and echinoderms have rigid internal plates of calcium carbonate. But the skeleton of vertebrates is different and quite remarkable. In some vertebrates, the skeleton is made of cartilage, a tough yet flexible protein-based tissue, but in most it is bone. Not only is bone surprisingly light and strong, making it very effective at supporting bulky bodies, but it has an extra and unexpected quality: it is alive. Intermixed in a matrix of protein and minerals are cells that deposit bone and others that remove bone. Other cells sense mechanical pressures and relay messages instructing the bone to grow or shrink and respond to the changing conditions. Bone is always dynamic. It is an extraordinary tissue and one ideally suited to large, active, growing animals, whether living in water or on land.



Vertebrates also differ from their closest invertebrate relatives, the tunicates and amphioxus, in their sophisticated brain and sense organs. The organization of the brain is very consistent across all the vertebrates, from lampreys to humans, with the three key sensory inputs being visual (paired eyes), chemical (paired olfactory sense organs), and mechanical (detecting pressure changes in water or sound in air). The entire head region of vertebrates is elaborate, based around a skull that encases the brain, yet presenting these sense organs to the outside world. The embryonic development of the skull reveals another oddity: a peculiar type of cell called the neural crest. These cells arise from the edges of the developing nerve cord and migrate through the embryonic tissues, before forming a whole variety of structures, including the bone or cartilage of the skull, jaws, and gill supports. Without neural crest cells, vertebrates could not build their complex, protected head region; without neural crest cells, they could never be the large predators and herbivores that dominate ecosystems on land and in the sea.

These features – large size, efficient blood circulation, dynamic skeleton, intricate brain, protective skull, and elaborate sense organs – combine to set vertebrates apart from their relatives. They may share the chordate phylum with amphioxus and tunicates, but the vertebrate body is much more complex and sophisticated. The differences within the phylum may run even deeper. Comparisons between the genome sequences from vertebrate and invertebrate animals have uncovered a fascinating fact. DNA sequences clearly show that early in vertebrate evolution, right at the base or very soon after, a major mutation occurred. The whole genome – every single gene – was doubled, and then doubled again. For every gene in an ancestral chordate, the early vertebrates had up to four. Some of these 'extra' genes were soon lost, but many were not, and consequently vertebrates have a greater diversity of genes than do most invertebrates. Whether new genes permitted the evolution of new vertebrate features is controversial, but one thing is certain: the invertebrate/vertebrate divide should not be ignored.



## The vertebrate tree

A common way to classify backboned animals is to divide them into fish, amphibians, reptiles, birds, and mammals. This works reasonably well for many purposes, but it does not accurately reflect the phylogenetic tree of vertebrates. One problem is that the various species of fish do not sit together on one single evolutionary line, separate from the other groups. Animals called 'fish' are found mixed in with other vertebrates. A similar problem applies to 'reptiles' since living reptiles share an evolutionary lineage with the birds. If we are to be strict in classifying animals according to evolutionary history, then groups called fish and reptiles should not exist.

Despite this complication, the path of vertebrate evolution, as known from fossils, molecular biology, and anatomy, is quite simple. The first vertebrates to evolve were fish-like in shape but lacked biting jaws. Although quite diverse in their heyday, over 400 million years ago, there are now only two surviving lineages of these jawless wonders: the lampreys and hagfish. The vertebrates with jaws evolved from jawless ancestors, and these early predators diversified into three main evolutionary lineages. These three groups are the 'chondrichthyans' (including sharks with their cartilaginous skeletons), the 'actinopterygians' (ray-finned fish), and 'sarcopterygians' (lobe-finned vertebrates). All three lineages include aquatic 'fish', but the sarcopterygians also include the vertebrates that left the water and emerged onto land. These animals, with strong skeletal structures in their fleshy fins, became the 'tetrapods' – vertebrates with four limbs. They comprise the amphibians, the 'reptiles', the birds (which perch on the same evolutionary twig as some reptiles), and the mammals.



## Lampreys and hagfish: surfeits and slime

With no biting jaws, the lampreys and hagfish need other ways to get food into their mouths. Adult lampreys have a sucker-like cup enclosing a circular rasping mouth armed with rings of sharp teeth. This ferocious-looking apparatus enables the animal to attach itself firmly to the flesh of its living prey, usually a large fish, and to suck its blood. Lampreys can stay fastened to their prey for several weeks, hanging limply as parasitic appendages.

When lampreys are attached by their sucker to prey, or even just to stones on a river bed, they cannot get oxygen from water taken in through the mouth. Instead, lampreys have 'tidal' gills: water is drawn into holes on the side of the animal's head, and expelled

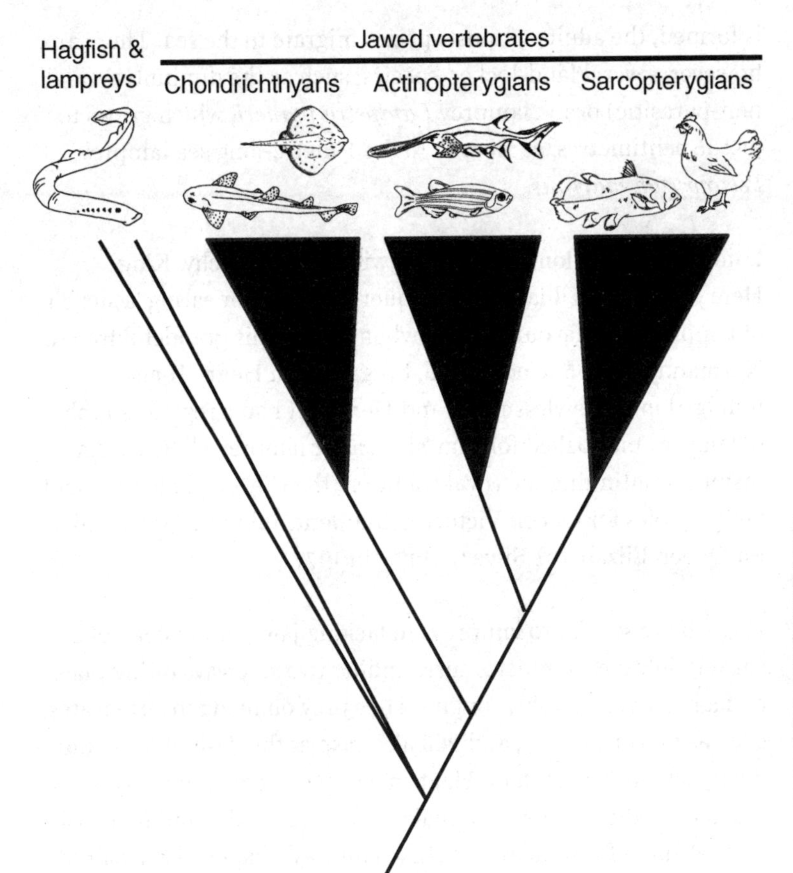


**12. Phylogenetic tree of vertebrates**

from the same. Larval lampreys, or ammocoetes, in contrast, have a normal unidirectional flow of water – in through the mouth, over the gills, and out through the gill slits – something made possible because juvenile lampreys are not parasitic. Lampreys spawn in shallow gravelly rivers, and after hatching the developing ammocoetes burrow into thick mud where they stay for several years, living on particles of food extracted from decaying matter. These slippery worm-shaped larvae can easily be found by digging in deep mud, whenever it is found close to fast shallows, in many British streams and rivers. After metamorphosis, when the sucker is formed, the adults of most species migrate to the sea. There are, however, several land-locked species, such as the diminutive (and non-parasitic) brook lamprey *Lampetra planeri*, which grows to just 15 centimetres, compared to the 1-metre-long sea lamprey *Petromyzon marinus*.

Lampreys have a long association with the monarchy. King Henry I, son of William the Conqueror, died after eating 'a surfeit of lampreys', his favourite dish, when visiting his grandchildren in Normandy in 1135. Undaunted, his grandson Henry II also indulged in the jawless treat, and Henry III had a regular supply of lamprey pies baked for him 'since after lamprey all fish seem insipid'. Continuing the royal tradition, the City of Gloucester sent lamprey pies for Queen Victoria's Diamond Jubilee in 1897 and for Queen Elizabeth's Silver Jubilee in 1977.

Hagfish are similar to lampreys in lacking jaws, but instead of a sucker, they have tentacles surrounding two sideways biting plates and a horny, retractable tongue. They prey on living invertebrates, such as marine worms, and will also rasp at the flesh of dead and dying fish on the sea floor. Hagfish will even gnaw their way into the dead bodies of larger animals, including whales and large fish, and eat them from the inside. Like lampreys, hagfish lack 'paired fins' located on the sides of the body; paired pectoral and pelvic fins are features characteristic of living jawed fish and, now transformed into legs, their land-dwelling descendants. The hagfish vertebral column is rudimentary compared even to that of lampreys, and for this reason, some zoologists do not even call hagfish 'vertebrates', using the term 'craniate' to encompass hagfish, lampreys, and jawed vertebrates. However, this view is contentious because features can be lost in evolution, and this might well be the case with hagfish vertebrae. We should not remove an animal from its natural group for reasons of secondarily losing something. In terms of overall anatomy and pattern of embryonic development, hagfish are really very similar to the rest of the vertebrates.

They do have a few peculiarities, however, and none more spectacular than slime. Many animals are slimy, but hagfish take it to a new level. Hagfish are the unchallenged masters of slime. When a hagfish is disturbed, pores along the sides of its body start to release a protein-based secretion that expands massively on contact with water. The quantity is impressive. Within seconds, a small 20-centimetre hagfish can produce several handfuls of thick, gluey slime, well suited to deterring predators. To avoid getting tangled in its own slime, the hagfish has a neat trick. It ties itself into a simple overhand knot which it then slides along its own body, wiping it clean.

## Jaws: sharks, skates, and rays



Everyone knows that sharks have jaws, whether fans of the 1975 movie or not. Jaws, together with paired fins, are the defining features of the three largest evolutionary groups of living vertebrates – the chondrichthyans (such as sharks and dogfish), actinopterygians, and sarcopterygians. In the words of Alfred Sherwood Romer, 'perhaps the greatest of all advances in vertebrate history was the development of jaws'. Studies on the embryos of dogfish, and other jawed vertebrates, have clearly shown how these highly efficient feeding structures evolved. In embryonic development, streams of migratory neural crest cells move down from the edges of the forming hindbrain into a series of bulges where they construct the skeletal supports for gills. In the jawed vertebrates – from sharks to humans – one of these bulges, the mandibular arch, develops not into a gill support, but into the bones or cartilage of the jaws. The stream of cells just behind it, the hyoid arch, forms a support structure connecting the back of the jaws to the skull. These pathways and patterns of cell migration in the embryo reveal that jaws almost certainly evolved from modified gill supports.

In living sharks, the upper jaw is not fused to the skull above but hangs quite loosely from elastic ligaments plus the hyoid-derived

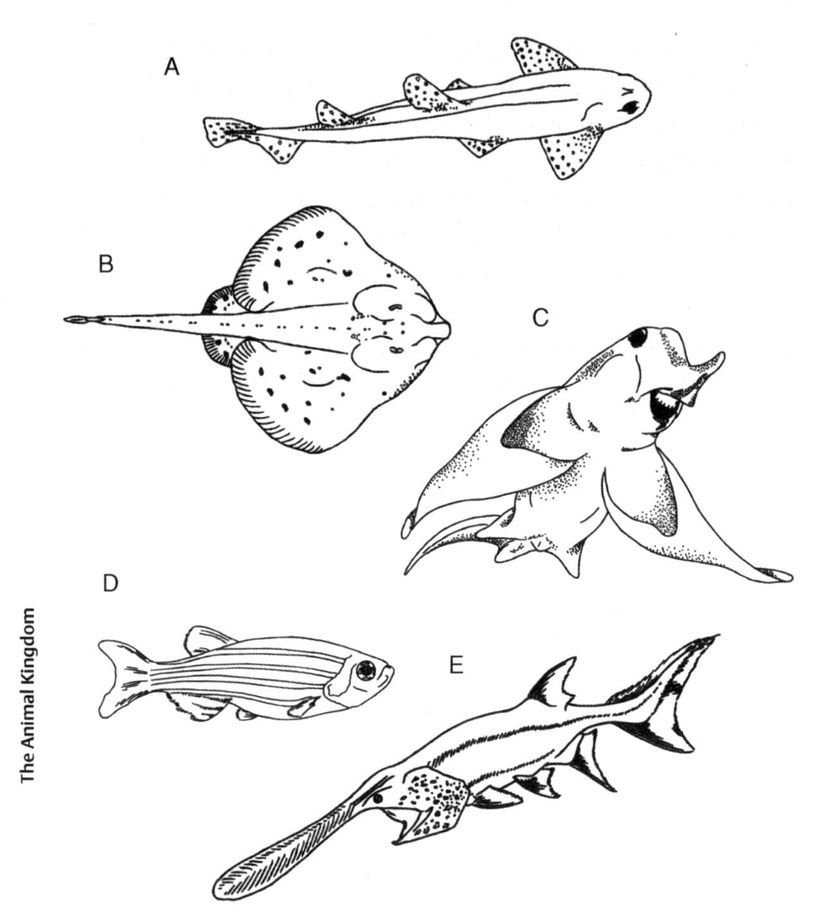


**13. A–C, Chondrichthyans: A, dogfish; B, little skate; C, chimera; D–E, Actinopterygians: D, zebrafish; E, paddlefish**

support at the back. This allows a feeding shark to protrude both jaws, either to pick small prey items delicately off the sea floor or to sink its teeth deeply into the flesh of large prey. The teeth of most sharks are sharply serrated and efficient at slicing through tissue once the jaws have been pressed deep into the prey, the cutting helped by the shark thrashing its body from side to side. Finding prey involves an impressive and sophisticated suite of sense organs. Sharks have a sense of smell that is both sensitive and directional. In some sharks, notably the Hammerhead

(*Sphyrna*), the two nostrils are placed far apart on strange extensions from the sides of the head, allowing them to deduce even more precisely the direction of highest chemical concentration. As they swim closer to their prey, sharks use both vision and mechanical sensing of vibrations in the water to home in. Just as they strike, many sharks slide a protective membrane over their eyes to prevent damage, which immediately reduces their vision. Temporary blindness does not give a chance for the prey to escape, however, as the shark then relies on exquisitely sensitive electroreceptors to detect the weak electric fields produced by animal muscles. These sensory cells are found in the 'ampullae of Lorenzini', a series of specialized pits in the shark's skin. They were first described in 1678 by Stefano Lorenzini, a brilliant Italian anatomist later imprisoned by the Grand Duke of Tuscany because of an alleged friendship with the Duke's estranged wife.

One notable character that sets sharks, dogfish, skates, and rays apart from most other jawed vertebrates is that their skeletons are made of cartilage not bone. They also lack a gas-filled cavity in their body; such a structure, called a swim bladder, is found in ray-finned fish and evolved into the lungs of land vertebrates. Without a swim bladder, it might be thought that sharks would sink to the bottom of the sea if they stopped swimming, but this is not the case. Instead, sharks solve the buoyancy problem in a quite different way. The key adaptation is a giant liver packed with oil, especially the long-chain hydrocarbon squalene, the low density of which counteracts the high density of the shark's skeleton, teeth, and scales, making sharks neutrally buoyant. Stability and additional lift is provided by stout paired fins on either side of the body. Buoyancy is also not a problem for most skates and rays, close relatives of the sharks, since these animals are usually benthic, meaning they live on the sea floor, in contrast to the pelagic, or free-swimming, sharks. Some giant rays, such as the manta ray, spend less time on the bottom and instead cruise through the ocean, flapping their greatly enlarged pectoral fins and straining plankton through a mesh of spongy tissue attached to their gill arches.

The final group of chondrichthyans, evolutionarily quite divergent from the sharks, dogfish, skates, and rays, are the strange ratfish, or chimeras. They too have a cartilaginous skeleton and lack a swim bladder, and they also have internal fertilization like sharks. But chimeras differ from other cartilaginous fish in that the upper jaw is fused to the skull and there is only a single gill opening on each side instead of several. The solidly built head with a fleshy elephant-like snout gives these animals a freakish appearance. They have an overall 'fishy' shape with large fins, but their big eyes and buckteeth best resemble a cartoon rabbit, while some species also have a long, rat-like tail. Appropriately, the name 'chimera' recalls the mythological monster of ancient Greece, made from components of various animals, described by Homer in the *Iliad* as 'a thing of immortal make, not human, lion-fronted and snake behind, a goat in the middle'.



## The ray-finned fish: flexibility

Most of the well-known species of fish belong to a diverse group called the actinopterygians, or ray-finned fish. Examples include commercial fish such as cod, haddock, herring, tuna, and eels; species kept in aquaria such as goldfish, tetras, guppies, and catfish; most fish sought by anglers such as trout, carp, pike, roach, and bass; and many others, including minnows, sticklebacks, and gobies. It is easier to list the 'fish' that are not actinopterygians, since these are only the hagfish, lampreys, sharks, skates, rays, chimeras, coelacanths, and lungfish. There are over 24,000 species of ray-finned fish inhabiting the world's oceans, seas, rivers, and lakes.

Ray-finned fish have both 'unpaired' and 'paired' fins, just like sharks. The unpaired fins lie along the midline of the body, and comprise one or more dorsal fins along the back, a caudal fin at the tail, and an anal fin on the ventral side. There are also the two sets of paired fins: pectorals just behind the gills and pelvics further back. In ray-finned fish, as the name suggests, the fins are supported by thin bony rays, giving them great manoeuvrability. This is particularly important for the pectoral fins, which can be twisted and flexed, enabling fine control whether the fish is swimming, turning, or even remaining motionless in the water. During the evolution of the actinopterygians, fins have been modified in many different ways, and this was clearly one factor underlying the diversification of this group. To give some extreme examples, knifefish can swim slowly either forwards or backwards using ripples sent down an enormously enlarged anal fin, while flying fish have large wing-like pectoral fins enabling them to glide through air for up to 50 metres. Tuna fish out-sprint their prey with bursts of speed made possible by concentrating movement to the caudal fin and posterior body; sea horses lack a caudal fin entirely and swim slowly using undulations of the dorsal fin.

The bony rays supporting the fins also show variation. In some species, they serve a defensive function, such as the sharp and protruding spines in the dorsal fin of a perch or stickleback, and in a few cases, they can inject venom, for example in stonefish, weever fish, and lionfish. They can be also used as accessory organs for feeding, such as in gurnards and angler fish. Gurnards are bottom-living fish with elongated pectoral fin rays, equipped with a range of sensory receptors, used for 'walking' along the sea bed and feeling for prey. In angler fish, the first three spines of the dorsal fin are exceptionally long and fused to form a 'fishing rod' used to lure prey towards a gaping mouth.

Ray-finned fish have solved the problem of density in a different way to sharks. Just below the vertebral column is a gas-filled cavity, the swim bladder, which acts as an internal float to help maintain buoyancy. In some fish, such as carp and trout, the swim bladder is connected by a tube to the gut, allowing it to be filled by gulping air at the surface. In other fish, such as perch, the connection to the gut has been lost and the swim bladder is filled by a specialized gland which secretes gases absorbed from the blood. Many fresh-water fish, including members of the carp family, also use their swim bladder to enhance hearing, using modified spines on the vertebrae to transmit vibrations of the gas bladder to the inner ear. Perhaps surprisingly, some fish can also make sounds with their swim bladder, used for attracting a mate or deterring rivals. For example, male toadfish (family Batrachoididae) make sounds by contracting fast 'sonic muscles' attached to the swim bladder, causing its walls to vibrate rapidly. This produces a sound like a loud, plaintive foghorn.

The head of ray-finned fish is complex and intricate. In most ray-finned fish, the left and right lower jaws can swing sideways, while one of the upper jaw bones, the premaxilla, can protrude forwards. These movements allow the mouth cavity to be enlarged suddenly, creating a strong suction force, used to capture prey items that would otherwise escape. Suction feeding is seen in a huge number of ray-finned fish and underpins the ecology of many species. At the back of the head are the gills. These are hidden behind a flap, the operculum, which not only protects the delicate gills but also plays a key role in how they function. By closing the operculum and widening the mouth, then shutting the mouth and opening the operculum, ray-finned fish effectively pump water across their gills, even when not swimming. Coupled with a blood supply to the gill filaments oriented in the opposite direction to the water flow, this allows ray-finned fish to extract maximal oxygen from water.

The great diversity of ray-finned fish is explained only partly by the invention of the swim bladder, the adaptability of fin rays, suction feeding, and the operculum. High species numbers are dependent on the integration of many factors, including a combination of ecological opportunities, an adaptable body plan, and possibly even features of the genome. On the last point, it is interesting that the teleost fish, which constitute the majority of ray-finned fish, share an additional genome duplication on top of the two that occurred at the base of vertebrates. It is currently unclear if this permitted greater adaptability of the body plan, or if it even caused enhanced speciation rates through different genes being lost in different populations. The extra genome duplication did not affect all ray-finned fish, and there are still a few living descendants from the earliest evolutionary radiation of this group: the so-called 'non-teleost actinopterygians'. These include the peculiar filter-feeding paddlefish (*Polyodon*) with its spatula-shaped head, the heavily armoured gars (*Lepisosteus*), and the various species of sturgeon, many now rare and endangered, whose eggs are prized as caviar.

# Chapter 10
# Deuterostomes III: vertebrates on land

Eye of newt, and toe of frog,
Wool of bat, and tongue of dog,
Adder's fork, and blind-worm's sting,
Lizard's leg, and howlet's wing,
For a charm of powerful trouble,
Like a hell-broth boil and bubble.

William Shakespeare, *Macbeth*, Act IV, Scene I

## From lobe-fins to legs

On 22 December 1938, a young museum curator in South Africa was shown an unusual, iridescent blue fish among the catch of a local fishing boat. The fish, almost 2 metres in length, with strong fleshy fins and heavily armoured scales, became a *cause célèbre*. It was the first living specimen described of a coelacanth, a member of a group of ancient fish with a fossil record dating from 400 million years ago until their supposed extinction 65 million years ago. The *London Illustrated News* described the discovery as 'one of the most amazing events in the realm of natural history in the twentieth century'. Specimens of *Latimeria chalumnae*, named after museum curator Marjorie Courtenay-Latimer, have since been caught many times off the East African coast, particularly near the Comoro Islands, and a second coelacanth species, *Latimeria menadoensis*, has been discovered in the Indian Ocean.

The excitement about living coelacanths is not simply that they were once thought extinct. More importantly, these are animals with special significance for understanding the evolution of land-dwelling vertebrates, a critical step in our own evolutionary history. The fleshy fins, which can be moved independently on the left and right sides as if the coelacanth is 'walking' in the open sea, are central to the argument. Their structure, together with various skull features, reveal that coelacanths belong to the 'sarcopterygian' group (lobe-finned vertebrates), and not to the ray-finned fish. In addition to coelacanths, the other two groups of living lobe-finned vertebrates are the lungfish and the tetrapods, the latter including all the land vertebrates including humans. Neither coelacanths nor lungfish are the actual ancestors of land vertebrates, but the three groups are related and they each descend from lobe-finned fish that swam in the early Devonian period, around 400 million years ago. Fossil evidence and molecular data suggest that tetrapods are evolutionarily slightly closer to lungfish than to coelacanths, but both groups of lobe-finned fish are important for understanding our origins. The living lungfish, of which there are four species in Africa, one in South America, and one in Australia, are all quite unusual and specialized animals, but they are indeed air-breathing fish – they have lungs equivalent to the lungs of land vertebrates.

While several groups of invertebrates, such as insects, myriapods, spiders, and snails, made the difficult transition from living in water to living on land, the same transition was only successful once in the entire evolution of vertebrates. The single evolutionary lineage of vertebrates that overcame the challenges of living on dry land gave rise to all of the land vertebrates still living today: all amphibians, all reptiles, all birds, and all mammals. To live successfully on land, animals must be able to obtain oxygen from air, find and catch food on land, carry their body weight in a medium far less supportive than water, propel themselves along land, and avoid drying out through excessive loss of moisture. Lungfish, close relatives of land vertebrates, can use their lungs to breathe air, as well as their gills to get oxygen from water, suggesting that air-breathing evolved long before the true transition to life on land. But supporting the body, feeding, and moving on land seem more of a challenge, and they demand several evolutionary changes between 'fish' and 'tetrapods'. Some remarkable fossils have cast light on these modifications, and have even revealed the order in which they took place.

One anatomical change was the evolution of a flattened snout capable of snapping at prey, rather than using the suction method that works so well underwater. Fossils of the extinct *Panderichthys* and *Tiktaalik*, which lived around 375 million years ago, show this feature well and must have given these animals a somewhat crocodile-like front end. However, they were still very 'fishy' in that their fins had delicate rays at the end of skeletal elements, rather than strong, bony fingers. This combination of fish-like and tetrapod-like features led Neil Shubin, the discoverer of *Tiktaalik*, to nickname this animal the 'fishapod'. *Acanthostega*, which lived a little later at 365 million years ago, had fins that ended in jointed finger-like elements, making it even more tetrapod-like. Interestingly, there were not just five fingers, as seen today in most living land vertebrates, but eight on the forelimb and probably the same number on the hindlimb. *Acanthostega* almost certainly lived in water and breathed using gills, but it may have been capable of forays onto land, perhaps to catch food or to bask in the sun. Another fossilized early tetrapod, *Ichthyostega*, may represent yet another step in the transition to living on land, since in addition to the above features, this animal had a stronger 'axial' skeleton or backbone, with longer bony projections, or 'zygapophyses', enabling vertebrae to interlock and help support the animal's body weight.



## Frogs and salamanders: the skin-breathers

Long-extinct animals from the Devonian period may represent the first tentative forays of backboned animals onto the land, but these animals would have still been heavily dependent on water, not least for reproduction. The latter also holds true for some tetrapods living today, a group of animals that spend much of their life on land, but lay their eggs in or around water. These are the frogs and toads, newts and salamanders, and the legless caecilians: the living amphibians. Most of these animals never stray far from damp habitats, since their skin is not particularly waterproof and, in many species, must be kept moist as a surface for gas exchange. A second reason for reliance on water is that their eggs and young demand a wet habitat. The larvae of most living amphibians, such as the tadpoles of frogs, even have external gills to extract oxygen directly from their aquatic habitat. When discussing amphibians, it is tempting – but inaccurate – to consider the living species as representing just another small stepping stone on the route to 'true' life on land, less successful and less advanced than the reptiles, birds, and mammals. But the fact that they exist at all today is evidence of their continuing success, and in reality, the living amphibians are greatly specialized and very different from the earliest land vertebrates. Furthermore, some individual species are quite numerous, particularly several species of frogs and toads. For example, the Cane Toad *Rhinella marinus* has spread and become so abundant in northern Australia, following its deliberate but disastrous introduction in 1935, that it is now a major invasive pest.

A few species of amphibian spend their entire lives in water, never venturing onto land at all, even as adults. Examples include the Japanese giant salamander *Andrias japonicas*, which can grow to 1.5 metres, the grotesque American 'Hellbender' salamander *Cryptobranchus alleganiensis*, and the African clawed toad, or *Xenopus*. But perhaps the best-known 'aquatic amphibian' is the Mexican axolotl, *Amblystoma mexicanum*, which resembles a large, 20-centimetre-long, sexually mature tadpole, complete with feathery external gills. That is exactly what it is, since axolotls evolved from 'normal' terrestrial salamanders through a change in their developmental physiology, and now become mature without

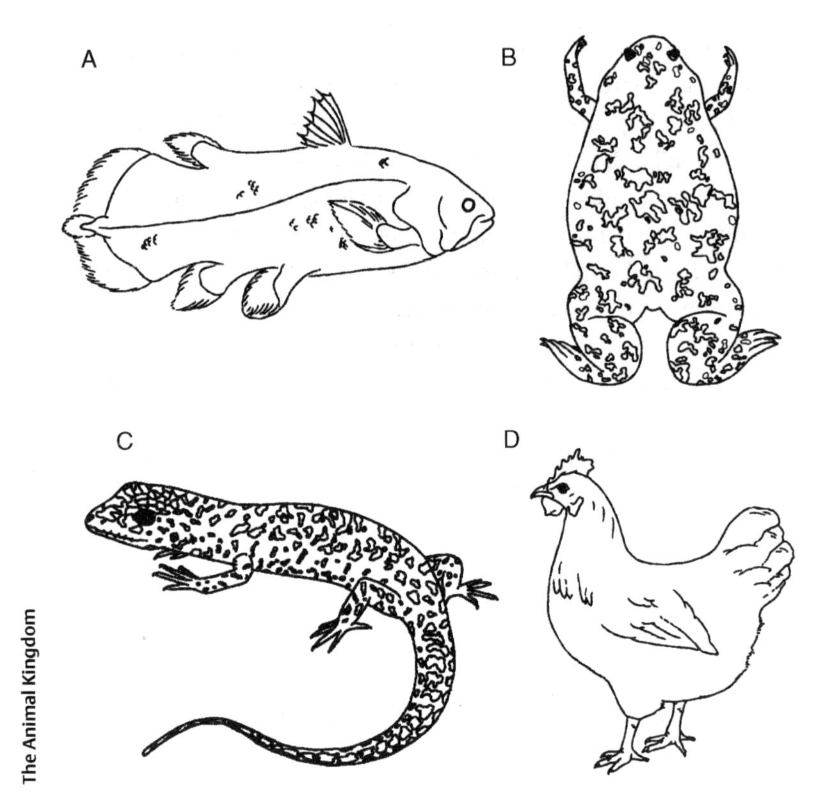


**14. Sarcopterygians: A, coelacanth; B, African clawed toad; C, Tasmanian snow skink; D, chicken**

going through metamorphosis into the ancestral adult form. The axolotl is a powerful reminder that evolution is not a one-way street and that different lineages of animals each adapt to their local conditions, irrespective of any overarching trends that we may perceive.

## Scales and sex: the reptiles

The reptiles represent a 'grade' of organization, rather than a single group on the evolutionary tree of vertebrates, and the living species comprise a disparate assemblage of animals including lizards, snakes, turtles, crocodiles, and the archaic tuatara of New Zealand.

Dinosaurs were also reptiles, on the same evolutionary line as crocodiles and birds, while other extinct reptiles included the winged pterosaurs and the marine ichthyosaurs, mosasaurs, and plesiosaurs. These aquatic species, like the marine turtles of today, went back to water secondarily – they evolved from ancestral species that lived fully on land. The defining feature of the earliest reptiles is that they marked a true break from aquatic habitats: the first group of vertebrates to do so. The reptiles that inhabit land can live, feed, and breed without ever returning to water.

Two key innovations seem to have been central to this transition – the evolution of a thoroughly waterproof skin and the invention of a shelled egg with several internal membranes. The first property seems clear enough, and was achieved by the evolution of more complex skin with several layers of cells producing keratin proteins and lipids. One implication of this change is that the skin cannot be used for respiration (as it is in modern frogs and salamanders) because only wet surfaces allow oxygen and carbon dioxide to diffuse across. Instead, reptiles evolved 'costal respiration' in which muscles attached to the ribs are used to ventilate the lungs, turning the lungs into even more effective breathing organs. The importance of the 'amniotic egg' is less obvious, but was also vital. The key lies in three membranes, the amnion, allantois, and chorion, which envelop the embryo and provide a large expanse of blood vessels for gas exchange, plus a site where toxic nitrogenous waste can be accumulated safely away from the developing body. Although most reptile species, including turtles and crocodiles, lay amniotic eggs encased within a shell, some snakes and lizards have live birth. Most commonly, as in garter snakes, boas, and vipers, this involves the mother retaining the large yolky eggs inside her body throughout their development. Some other reptiles provide nourishment directly from the mother, rather than from yolk; in the most extreme cases, this can involve a placenta, as in the *Mabuya* and *Pseudemoia* skinks. The many physiological, anatomical, and behavioural adaptations of reptiles have allowed them to invade some of the hottest and driest environments on the planet, including the baking deserts of Africa, Australasia, Asia, and the Americas.

The body physiology of reptiles is well suited to warm conditions, since most bask in the sun's rays to increase their body temperature. This enables a high metabolic rate and active lifestyle to be sustained even without good insulation of the body. Temperature also affects the biology of many reptiles in a very different and rather unusual way: it can determine the sex of their offspring. For example, if the eggs of an American alligator are incubated below 30°C, they hatch into females, while eggs kept at 33°C develop as males. This phenomenon, called 'temperature-dependent sex determination', or TSD, contrasts with the more familiar 'genotypic sex determination' in which genetic differences control the sex of offspring, such as the male-determining genes found on the Y chromosome of mammals. But why should some reptiles (and, for that matter, some fish) use TSD, when at first sight the gene-based method seems more reliable? Is there not a danger that a shift in environmental conditions, such as a changing climate, could drive a TSD population to extinction, since every offspring might then develop as the same sex? The answer seems to lie in adaptation to local ecological conditions, as demonstrated neatly by recent research on the Tasmanian snow skink, *Niveoscincus ocellatus*, undertaken by Ido Pen, Tobias Uller, and colleagues. This reptile lives from sea level up to mountainous regions, and – remarkably – populations at low altitude have TSD, while animals of the same species at high altitude use a genotypic method. The reason for the difference seems to be that mothers at low altitude use TSD to produce proportionately more daughters in warm years which then have maximal opportunity to grow large and fecund during the longer summers, but they switch to producing more sons in cooler years because male size is less important in snow skinks. This advantage is lost at high altitudes where growth rates are slower and where the more wildly fluctuating temperatures would play havoc with the ratio between the sexes unless TSD gave way to genotypic sex determination.

## Feathers and flight: the birds

One celebrated group of reptiles, the dinosaurs, dominated terrestrial life on Earth for many millions of years. The first dinosaurs evolved around 230 million years ago and diversified into a multitude of species of different sizes, shapes, and habits, until their sudden and famed extinction 65 million years ago. Or to be more accurate, until their apparent extinction. The popular notion of a total wipe-out of dinosaurs is a little deceptive because some animals alive today are direct evolutionary descendants of a group of dinosaurs, the theropods. Well-known extinct theropods include the giant carnivore *Tyrannosaurus* and the smaller, but perhaps equally fearsome, *Velociraptor*, made famous in the film *Jurassic Park*. Of course, *Tyrannosaurus* and *Velociraptor* are no longer roaming the Earth, but you see some of their close relatives every day. One group of theropods did not go extinct 65 million years ago, but survived the global catastrophe and diversified until the present day. They are the birds.

From an evolutionary perspective, birds are a group of dinosaurs that did not become extinct. The idea that birds evolved from dinosaurs was first proposed by Thomas Henry Huxley in the 1860s. Huxley noted key similarities between the layout of the skeleton in theropod dinosaurs and in an extinct bird, *Archaeopteryx*, known from a few beautifully preserved 150-million-year-old fossils. Although *Archaeopteryx* had features that were very lizard-like, such as teeth and a long bony tail, it also had wings and feathers. Then, as now, *Archaeopteryx* was considered to be one of the first birds to evolve. Huxley's views were contentious, and although every biologist accepted that birds evolved from ancient reptiles, the idea that they were actually direct descendants of dinosaurs soon fell out of favour. The idea lay dormant for most of the 20th century until thrust back into the mainstream in the 1970s through the careful work of John Ostrom at Yale University. But the most dramatic and clinching piece of evidence did not come until the 1990s, when several remarkable fossils of 'feathered dinosaurs' were discovered in China – these were undisputed 'non-flying' dinosaurs, but with feathers covering their body and legs. Not only do the feathered dinosaurs provide strong evidence for a bird–dinosaur relationship, they also highlight feathers as an early adaptation, possibly for keeping warm, that paved the way for the later origin of flight.

The feathers of modern birds are remarkable structures. Feathers used in flight have an intricate and asymmetrical structure that provides rigidity and power on the downstroke, while also being strong and incredibly light. They have a central shaft, or rachis, from which protrude a myriad closely spaced barbs, each bearing minute hooked barbules that interlock. In contrast, the 'down feathers' used for body insulation do not interlock in the same way, and so trap pockets of air rather than producing sheet-like surfaces. As well as their two principal functions – flight and insulation against cold – feathers play roles in water-proofing, camouflage, and communication between individuals. Feathers and flight dominate the entire ecology and behaviour of birds, and have combined to shape their evolution. Weight is an important issue in flight, and accordingly birds have evolved to have quite thin, hollow bones, strengthened by internal struts. Heavy teeth have been lost in evolution, as has the long tail. But more important than absolute weight is its distribution, and so the anatomy of birds is adapted to place the centre of gravity further forward than in most vertebrates, directly between the wings. This has been achieved by tucking the 'thighs' of the hind legs forward along the sides of the body and lengthening the foot; this explains why birds' knees seem to point backwards – they are not actually knees but ankles.

There are around 10,000 species of bird alive today, found living on every continent and flying over every sea. They include tiny hummingbirds in South America, spectacular birds of paradise in the forests of New Guinea, majestic condors soaring through Andean passes, shearwaters skimming over ocean waves hundreds of miles from land, kestrels hovering over grassy verges, secretive wrens, robins, thrushes, and more. This may sound like a picture of diversity, but in reality all birds are very much alike – at least in terms of anatomy. The most striking exceptions are the few birds that have secondarily lost the ability to fly. Penguins with their unusual body form are adapted to water not air, and ostriches with their great size and bulk are flightless: these exceptions serve to remind us that flight places massive constraints on the anatomy and physiology of birds. Evolution cannot circumvent the laws of physics.



## Milk and hair: the mammals

The active lifestyle of birds is only possible because of their relatively warm body temperature, generated by a high metabolic rate coupled with the insulation provided by feathers. The other land vertebrates that generate and maintain their own body heat are the group to which we belong: the mammals. In the case of mammals, however, the vital insulation is provided by hair. The structure of hair is far less complex than that of feathers, comprising rather simple strands made from fibres of the protein alpha-keratin. Even so, overlapping layers of hair can trap air very effectively and so keep warmth in. This retention of body heat allows mammals to get out and about in cold conditions, before the sun's rays have had a chance to warm their reptilian relatives. Unlike birds, mammals did not evolve from within the familiar diversity of reptiles. In an evolutionary tree of amniotes (the land vertebrates with an amniotic egg), one division gave rise to lizards, snakes, crocodiles, dinosaurs, and birds, while a sister lineage – the extinct synapsids – gave rise to the mammals.

Besides hair, a second important character shared by all mammals is lactation: the production of milk to provision offspring. This is a crucially important adaptation since it allows mammals to reproduce at any time of year, even when a diversity of food is not easily found or its availability fluctuates in time. Adult females can stock up on food whenever it is available and store energy as fat reserves; offspring are then provisioned with high-energy milk by suckling from their mother. Food-gathering by an experienced adult is also likely to be more efficient than by a juvenile, meaning that suckling on milk allows the infant to put a greater proportion of energy towards growth.

Strange as it may sound, the use of milk may have also paved the way for the great ecological diversity of mammals, through the evolution of complex teeth. The argument goes as follows. Because of lactation, newborn mammals do not need teeth. This means that the skull and jaw grow substantially before teeth arise; in turn, this enables a shift away from continual replacement of simple teeth, the system seen in most amniotes including lizards. Instead, mammals evolved 'diphyodonty', meaning that just two sets of teeth are produced: one simple set in juveniles and then complex teeth in the almost full-size jaw. Because tooth development is delayed in this way, mammalian teeth could evolve to have precise matches between the upper and lower jaws, a feature known as occlusion. Such a property is hard to envisage in an animal whose jaw grows extensively while containing teeth. Occlusion gave mammals the crucial ability to chew and grind food, especially tough plant matter, or to slice meat cleanly off their prey. Equipped with such formidable apparatus, the early mammals diversified to exploit a greater range of food sources and feeding strategies than seen in any other group of vertebrates.

There are about 4,400 species of mammals, less than half the number of birds, yet they display a greater diversity of body shapes, sizes, and modes of life. Of these, there are just five living species of monotreme, or egg-laying, mammal: the platypus, and four types of echidna or spiny anteater. All other mammals are 'therians' and have live birth. These include a few hundred species of marsupials which give birth to very immature infants and nourish them within a pouch, such as kangaroos, wombats, opossums, potoroos, bandicoots, koala, and the Tasmanian devil. The vast majority of living mammals are placentals, which have longer pregnancies and no pouch. The ecological diversity of placentals is staggering, and includes insectivores such as shrews, grazing herbivores like antelope, elephants, giraffe, and bison, hunting predators such as foxes and lions, opportunist omnivores such as mice, rats, and humans, aquatic herbivores such as manatees, aquatic predators like seals and dolphins, and even a group of mammals that took to the skies, the bats.

For most of the 20th century, there has been confusion over the true phylogeny of placental mammals. Amongst all this diversity, who is most closely related to whom? The question is now much closer to resolution, particularly since the recent application of DNA sequencing technologies. The emerging consensus divides the placental mammals into four great lineages. Remarkably, these lineages map beautifully onto the known geological history of the continents, suggesting that diversification of the placentals occurred just as the major landmasses of the world were separating. There is the 'Afrotheria', which, as the name suggests, comprises the mammalian orders that originated in Africa, including elephants, aardvark, and manatees. From the Americas came the 'Xenarthra', comprising anteaters, sloths, and armadillos. The Laurasiatheria includes a range of mammals thought to have evolved on the northern supercontinent of Laurasia, the forerunner of Europe plus much of Asia. These include cats, dogs, whales, bats, shrews, cows, and horses, amongst many others. Finally, the Euarchontoglires or Supraprimates includes rats, mice, and rabbits, plus primates such as monkeys and apes.

Standing back and looking at our own place in the evolutionary tree of animals, humans represent but a tiny twig. We sit within the primates, which in turn are within the Euarchontoglires.

These fit within the placentals, which are part of the therians, which lie nested inside the mammals, which are part of the amniotes, in turn within the tetrapods, and then inside the sarcopterygians. The sarcopterygians are one of the three groups of jawed vertebrates, within the vertebrates, inside the chordates, within the deuterostomes, within the bilaterians, inside the great tree of animal evolution.

# Chapter 11
# Enigmatic animals

> There are known knowns; there are things we know that we know. There are known unknowns; that is to say there are things that we now know we don't know. But there are also unknown unknowns, there are things we do not know we don't know.
>
> Donald H. Rumsfeld, US Department of Defense briefing, 2002

## New phyla, new insights

The history of zoology has been a story of changing opinions. There has been a century of debate and argument surrounding the evolutionary relationships between animals, and the problem is compounded by the fact that new species are discovered daily. For hundreds of species, more is learned each year about anatomy, ecology, development, and behaviour. It is pertinent, therefore, to stand back and ask how accurate is our current state of knowledge? Will there be further great overhauls, or do we now have a secure framework from which to delve deeper into animal biology? We must first ask whether we really know the full diversity of the Animal Kingdom.

It is a certainty that many thousands, or even millions, of animal species remain to be discovered. Tropical rainforests and the deep sea are just two ecosystems that teem with life, yet where scientific exploration has just scratched the surface. However, finding a new species, or even a thousand new species, does not radically change our understanding of animal biology. It is certainly significant in other ways; for example, knowing all species in an ecosystem can help in attempts to understand nutrient cycling and patterns of energy flow. Such insights are important. But most new species that are discovered are close relatives of already known species and so, if we wish to comprehend the overall pattern of animal diversity on the planet, such findings are not the key. They add detail, but they do not force a radical change to our state of knowledge.

The story is different at higher taxonomic levels. The most fundamental category in the classification of animals is, of course, the phylum. To use Valentine's words again, 'phyla are morphologically-based branches of the tree of life'. Discovery of a new phylum, therefore, really does change our state of knowledge. It adds a new branch to the animal evolutionary tree, and just as importantly reveals a new morphology: a new way to build the body. Putting the two together – a new branch and a new morphology – in turn can change our views on when, why, and how particular characters arose in evolution: perhaps fundamental characters such as symmetry, segmentation, or a central nervous system. But are there any phyla left to be discovered?

In this book, I recognize 33 different animal phyla, and most of these have been known for a very long time. By the late 20th century, many zoologists thought that all phyla must have been discovered. There was a surprise, then, in 1983 when the Danish zoologist Reinhardt Kristensen described a new species that was so different from everything else that it needed an entirely new phylum. He named this phylum the Loricifera. These minute animals, usually far less than a millimetre in length, look like miniature urns or ice-cream cones, clinging to sand grains. A few other zoologists had noticed these animals in the 1970s, including Robert Higgins, after whom the loriciferans' swimming Higgins larva is now named. But the biggest surprise is that the new phylum was not found in a remote and inaccessible part of the world, but just off the coast of Roscoff in France, home to a busy marine biology research centre.

The next 'new phylum' to be discovered – the Cycliophora – was again simply being overlooked. Cycliophorans are tiny symbiotic animals that live on the mouthparts of scampi (*Nephrops*) and lobsters (*Homarus*). These host species are so common that thousands of people must have eaten a rogue cycliophoran without ever knowing what zoological curiosity they were consuming. Remarkably, it was Kristensen again – together with Peter Funch – who described the new animal in 1995, after it had been first spotted by Tom Fenchel.

A third new body plan was reported in 2000, and this time it really was from a remote field site, visited by very few scientists. It had been found when Kristensen (again) was leading a field trip of students to Disko Island off the coast of Greenland. There, the students discovered some unusual microscopic animals living in an icy freshwater spring. Just one-tenth to one-eighth of a millimetre in length, with complex jaws that could be projected out of the mouth, their anatomy was so different from anything else that they deserved at least a new class, and possibly a new phylum. They were named the Micrognathozoa, meaning 'miniature jawed animals'.

So could there be yet another phylum waiting to be discovered? Quite possibly. The three examples above are all minute animals, far less than a millimetre in length, and it is amongst such microscopic fauna that a similar future discovery might be made. A promising place to look would probably be amongst the 'meiofauna', the animals that live between grains of sand. The discovery of the Loricifera off the French coast implies that such a finding could be made anywhere in the world, yet even so, I would suggest that remote deep-sea habitats might be the best bet. But if you really want to describe a new phylum of animals, I would advise that you don't start with hunting for a new species. Instead, a new phylum might well be present amongst the old.

## New phyla from old?

It may sound paradoxical, but many changes to the list of known animal phyla, whether they are discoveries of new phyla or mergers between old phyla, have come about because of more detailed studies on previously described species. A phylum should contain animals from one evolutionary branch; hence, if new data indicate that a phylum contains superficially similar species from different parts of the evolutionary tree, then that phylum must be split in two. There is no alternative. This has happened several times in the past two decades, particularly as DNA sequence data have been used to test the evolutionary relationships between animals. When the DNA data highlight a glaring oddity – a species out of place in the tree – then classification has to change. A new phylum can be made for an old species.



The most important, yet still controversial, examples concern two groups of unusual worms. These are the Acoelomorpha and Xenoturbellida. Although neither contains very familiar animals, or even common animals, they have been known to science for a long time. So while the species are not new, they may still merit one or two new phyla. The Acoelomorpha contains small, marine, flatworm-like creatures, usually a few millimetres in length. One of the easiest to find is the beautiful *Symsagittifera roscoffensis*, or 'mint sauce worm', which is bright green in colour because of algae living inside its body. The worm lives on sandy beaches around Europe, especially on the French coast near Roscoff, where it can be found as 'slicks' of greenish sludge in wet puddles. If you creep slowly towards the sludge, it has the disconcerting habit of disappearing; this is a living sludge made of thousands of green worms that simply crawl into the sand when disturbed. These worms, and many like them, were traditionally placed in the phylum Platyhelminthes, alongside the 'true' flatworms, flukes, and tapeworms. There were always a few dissenting voices calling attention to unusual features of their anatomy, but for the most part, the mint sauce worm and its allies stayed within the Platyhelminthes. Only when gene sequences were compared did it become abundantly clear that they were not at all closely related to flatworms, flukes, and tapeworms, and a new phylum was proposed.

It was a similar story for the Xenoturbellida. These animals are much larger than the acoelomorphs, with the first species *Xenoturbella bocki* from a Swedish fjord being several centimetres in length and a recently discovered Pacific species even bigger. They are also flattened worms and not very impressive to look at, being rather simple, yellowish-brown animals with few discernible organs apart from a blind-ending gut. They too were considered to be platyhelminths by most zoologists, although some argued for closer relationships to echinoderms or hemichordates. One suggestion, based on initial DNA sequence analyses, was that *Xenoturbella* was a mollusc, but this conclusion was shown to be an unfortunate error caused by extracting DNA from *Xenoturbella*'s last meal rather than from its own cells. When enough genuine *Xenoturbella* DNA was extracted, and many gene sequences analysed, it became clear that the animal is not a platyhelminth, nor a mollusc, echinoderm, or hemichordate, but something quite distinct from other animal groups. A new phylum was proposed in 2006.



It seems possible that a few more animal phyla might be found lurking as already known animals, misclassified into the wrong part of animal phylogeny. So where might one look? There are dozens of unusual invertebrate animals that share only some characters with their supposed relatives. The challenge for zoologists is to determine which of these represent simply aberrant members of their phylum – where evolution has modified the body plan – and which ones have been misleading zoologists for decades. For example, Katrine Worsaae has drawn attention to the unusual marine worm *Diurodrilus*, presently considered an annelid but possessing few of the normal annelid characters and possibly even lacking segmentation. *Lobatocerebrum* is a similar case, as this worm has characters of both annelids and platyhelminths. Myzostomids, unusual annelid-like parasites of sea lilies, are another problem group. Might any of these be a new phylum?

Another oddity, and possibly the most bizarre animal on the planet, is *Polypodium hydriforme*. This tiny animal spends most of its life actually inside the eggs of sturgeon, better known as caviar, and when it emerges, it breaks up into a swarm of microscopic jellyfish. It may actually be related to jellyfish and a member of the phylum Cnidaria, but if so, it is certainly a strange one. It might be related to *Buddenbrockia plumatellae*, a weird worm-shaped parasite without a clear front, back, top, bottom, left, or right, and with no central nervous system. Both animals possess structures rather like the stinging capsules of Cnidaria. Molecular analyses suggest that *Buddenbrockia* does indeed belong to the Cnidaria, which means that the phylum to which it was formerly assigned, Myxozoa, must be subsumed within Cnidaria. New data can therefore remove phyla from the list, as well as generate new ones.



## The view ahead

Why does it matter whether some of these unusual animals fall into phyla of their own? The key reason is that every time we place a particular body plan, or unique morphology, onto the phylogenetic tree of animal life, it changes our perspective about the path of evolution. Consider the Xenoturbellida and Acoelomorpha. Animals in each of these groups have bilateral symmetry, but they lack a major central nerve cord in the midline of the body. This is, of course, in contrast to the situation in the majority of Bilateria – most ecdysozoans, lophotrochozoans, and deuterostomes have a main nerve cord. If the centralized nerve cord is a general feature of bilaterian animals, perhaps these two new phyla descended from very early branches in the evolutionary tree of animals? Did Xenoturbellida and Acoelomorpha, or perhaps just one of them, branch off before the divergence of Ecdysozoa, Lophotrochozoa, and Deuterostomia (but after Cnidaria)? If so, perhaps they provide us with a tantalizing glimpse of how the first bilaterally symmetrical animal bodies could have functioned, before the evolution of a main nerve cord for integrating information. Initial molecular analyses suggested this was indeed the case, at least for Acoelomorpha, although the conclusion was controversial. An alternative molecular study places Acoelomorpha and Xenoturbellida among the deuterostomes, alongside the echinoderms, hemichordates, and chordates. If this is correct, why do they not have a main nerve cord? Have they lost it in evolution, by spreading the nervous system around the body? Or are we wrong in our views about the common ancestor of Bilateria? These are important questions to address, but they hinge on pinning down exactly where Xenoturbellida and Acoelomorpha fit in the tree of life, something that has proved surprisingly difficult even with large amounts of molecular data.



This controversy brings us onto whether we should have confidence in the current phylogenetic tree of the Animal Kingdom. The 'new phylogeny' sees some early diverging non-bilaterian lineages (Porifera, Placozoa, Ctenophora, Cnidaria) separating from the branch leading to Bilateria, which in turn splits into three great superphyla: Ecdysozoa, Lophotrochozoa, and Deuterostomia. How sure can we be of this scenario? Hypotheses of evolutionary relationships have changed drastically over the past century, so might they change again? I predict they will not. Instead, I argue that it is time to have confidence in the 'new animal phylogeny', at least in broad outline. The phylogeny is based almost entirely on comparison of DNA sequences from genes found in all animals. And while the earliest molecular trees were built from one or a few genes, the basic framework has since been corroborated by massive analyses involving over a hundred genes per species. DNA sequence provides a mine of information on past history and, although not straightforward to analyse, it has provided the most robust, and the most internally consistent, data set ever applied to these problems. It is true that a few animals, such as Acoelomorpha, have not proved easy to place even using molecular data, but at least these methods leave them as unplaced or controversially placed, not squeezed into the most convenient place.



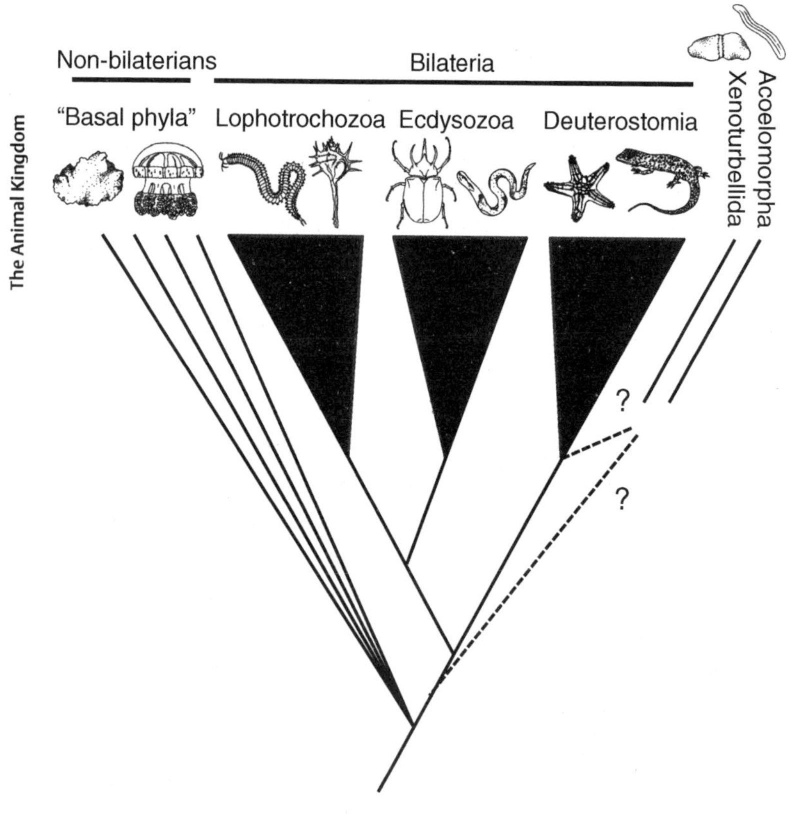


**15. Phylogeny of the Animal Kingdom showing alternative hypotheses for the position of Xenoturbellida and Acoelomorpha**

I believe we are at a time in the history of zoology when we have, for the first time, a robust evolutionary tree of animal diversity. We must remember, however, that this phylogenetic tree is just a starting point for biological investigation. A tree in itself does not provide understanding. What it provides is a framework that allows us to interpret biological data carefully and rigorously. Morphological studies, formerly used to build trees, now become more valuable than ever before as they can be interpreted in the light of an independent tree. Only with the robust framework of a phylogenetic tree can we compare anatomy, physiology, behaviour, ecology, and development between animal species in a meaningful way: a way that gives insight into the pattern and process of biological evolution.

# Further reading

## Chapter 1

L. W. Buss, *The Evolution of Individuality* (Princeton: Princeton University Press, 1987)

## Chapter 2

A. L. Panchen, *Classification, Evolution and the Nature of Biology* (Cambridge: Cambridge University Press, 1992)

J. A. Valentine, *On the Origin of Phyla* (Chicago: University of Chicago Press, 2004)

## Chapter 3

M. J. Telford and D. T. J. Littlewood (eds.), *Animal Evolution – Genomes, Fossils and Trees* (Oxford: Oxford University Press, 2009)

## Chapter 4

R. Dawkins, *The Ancestor's Tale* (Boston: Houghton Mifflin, 2004)

## Chapter 5

R. A. Raff, *The Shape of Life: Genes, Development, and the Evolution of Animal Form* (Chicago: University of Chicago Press, 1996)

S. B. Carroll, *Endless Forms Most Beautiful: The New Science of Evo Devo and the Making of the Animal Kingdom* (New York: W. W. Norton, 2005)

## Chapter 6

R. B. Clark, *Dynamics in Metazoan Evolution: The Origin of the Coelom and Segments* (Oxford: Clarendon Press, 1964)
J. A. Pechenik, *Biology of the Invertebrates*, 3rd edn. (New York: McGraw-Hill, 2009)

## Chapter 7

D. Grimaldi and M. Engel, *Evolution of the Insects* (Cambridge: Cambridge University Press, 2005)

## Chapter 8

H. Gee, *Before the Backbone* (London: Chapman & Hall, 1996)

## Chapter 9

J. A. Long, *The Rise of Fishes* (Baltimore: Johns Hopkins University Press, 2010)

## Chapter 10

F. H. Pough, C. M. Janis, and J. B. Heiser, *Vertebrate Life*, 5th edn. (New Jersey: Prentice Hall, 1999)

# “牛津通识读本”已出书目

古典哲学的趣味
人生的意义
文学理论入门
大众经济学
历史之源
设计，无处不在
生活中的心理学
政治的历史与边界
哲学的思与惑
资本主义
美国总统制
海德格尔
我们时代的伦理学
卡夫卡是谁
考古学的过去与未来
天文学简史
社会学的意识
康德
尼采
亚里士多德的世界
西方艺术新论
全球化面面观
简明逻辑学
法哲学：价值与事实
政治哲学与幸福根基
选择理论
后殖民主义与世界格局
福柯
缤纷的语言学
达达和超现实主义
佛学概论
维特根斯坦与哲学
科学哲学
印度哲学祛魅
克尔凯郭尔
科学革命
广告
数学
叔本华
笛卡尔
基督教神学
犹太人与犹太教
现代日本
罗兰·巴特
马基雅维里
全球经济史
进化
性存在
量子理论
牛顿新传
国际移民
哈贝马斯
医学伦理
黑格尔
地球
记忆
法律
中国文学
托克维尔
休谟
分子
法国大革命
民族主义
科幻作品
罗素
美国政党与选举
美国最高法院
纪录片
大萧条与罗斯福新政
领导力
无神论
罗马共和国
美国国会
民主
英格兰文学
现代主义
网络
自闭症
德里达
浪漫主义
批判理论

德国文学
戏剧
腐败
医事法
癌症
植物
法语文学
微观经济学
湖泊
拜占庭
司法心理学
发展
农业
特洛伊战争
巴比伦尼亚
河流
儿童心理学
时装
现代拉丁美洲文学
卢梭
隐私
电影音乐
抑郁症
传染病
希腊化时代
知识
环境伦理学
美国革命
元素周期表
人口学
社会心理学
动物
电影
俄罗斯文学
古典文学
大数据
洛克
幸福
免疫系统
银行学
景观设计学
神圣罗马帝国
大流行病
亚历山大大帝
气候
第二次世界大战
中世纪